Death by Technology

Death by Technology

The Road to Hell Is Paved with Good Inventions

JOHN R. COOK

McFarland & Company, Inc., Publishers
Jefferson, North Carolina

LIBRARY OF CONGRESS CATALOGUING-IN-PUBLICATION DATA

Names: Cook, John R., 1962– author.
Title: Death by technology : the road to Hell is paved with good inventions / John R. Cook.
Description: Jefferson, North Carolina : McFarland & Company, Inc., Publishers, 2021. | Includes bibliographical references and index.
Identifiers: LCCN 2020051068 | ISBN 9781476680309 (paperback : acid free paper) ∞ ISBN 9781476642277 (ebook)
Subjects: LCSH: Technology—Social aspects. | Technology—Environmental aspects. | Technology—Risk assessment. | Inventions—Social aspects.
Classification: LCC T14.5 .C6655 2021 | DDC 303.48/3—dc23
LC record available at https://lccn.loc.gov/2020051068

BRITISH LIBRARY CATALOGUING DATA ARE AVAILABLE

ISBN (print) 978-1-4766-8030-9
ISBN (ebook) 978-1-4766-4227-7

Printed in the United States of America

McFarland & Company, Inc., Publishers
Box 611, Jefferson, North Carolina 28640
www.mcfarlandpub.com

For future generations.
May you forgive us.

An Author's Note and Acknowledgments

One of the more interesting contrasts of the digital era is found in the common description of it as the age of information. Yet, at the same time it is an age of disinformation, and people seem to believe more erroneous information than ever before. Disputes rage in the digital world as to the source and veracity of information. Efforts to undermine the credibility of information necessitate greater efforts to present information in a verifiable manner. To that end I have endeavored to present reference material in a manner that the reader may use to verify any material and contentions made herein. The text is written in APA-style with in-text citations and an associated bibliography at the end.

This work is intended as a reasoned review grounded in scientific evidence and scholarly research. In service to this idea I have used primary sources for governmental data and reports from sources such as the United Nations, the World Health Organization, United States Government, the Atomic Energy Commission and other similar organizations. Studies cited throughout the text are from peer reviewed scientific journals whenever possible. For news items, publicly available information and agreed-upon versions of events sources such as the Associated Press, Reuters News Service, newspapers of record such as *The New York Times*, *The Washington Post*, *The Guardian*, and *The Atlanta Journal-Constitution* are utilized.

I would also like to thank Steve Platek, Andrew Kelly, Arjun Garg, and Charles Sugatapala for comments and discussions that have assisted in the development of the ideas in this text.

Table of Contents

Preface

For the vast majority of human existence, technology was minimal and consisted primarily of basic tools used for hunting or food preparation. The machine age ushered in an era of increasing reliance on technology that has culminated in the digital age, where technology has become part of the fabric of society. No longer are nation-states and the nuclear family the sole determinants of inputs into its citizens and members. No longer does the tribe control the myths and beliefs of its in-group. There has become a larger entity affecting the worldwide culture. That force is technology.

Technology is ubiquitous in modern life. Every aspect of existence from paying bills to social interaction has a technological overlay. Every problem is dictated to have a technological solution. This approach, sometimes referred to as "solutionism," suggests that every problem and every task should have a technological aspect. This occurs even when there is no evidence that a technological solution exists or is even appropriate.

Such an overreliance on technology suggests that there is more to this perpetuation of technology as the answer to all problems than a desire to solve problems. Instead, the emerging techno-culture is more akin to a religious movement than any other entity and has invaded every aspect of life.

Just as occupying forces impose their language, customs and beliefs on those they subjugate, so does technological society. Couching actions in terms of improving the lives of those so subjugated, the conquerors do not question whether forced conversions are desired. The people of South America largely speak Spanish and Portuguese and practice Catholicism not because of a collective choice but rather because these were imposed upon them. The rationale is always that it is done for the good of the converted.

With all religions there is a collective denial of the absurdities contained in certain central beliefs and promises. The ability to ignore contrary evidence or to twist it to suit the myths of the belief system is inherent in religious dogma. Technology as deity can be illustrated in one of the favorite allegories of the adherents of the faith. This story involves the proud proclamation that a teenager in Africa with a smart phone has access to

1

more information that the president of the United States did just twenty years ago.

While this may be true, the allegory also fails to acknowledge that this same individual may have no access to clean drinking water, experiences constant food insecurity, and in all probability does not have the means to charge the phone. It is a metaphorical prayer book in Latin for the Conquistadors to present to the newly conquered and converted.

This emerging culture is not one bounded by geopolitical lines on a map. It is not delineated by linguistic divisions or belief systems. It is not even completely visible. Instead the emerging culture, the über-culture of the planet—with its influence touching and affecting all other cultures—is one that is marked by access to, and acquisition of, the newest and most innovative technology. Those at the heart of the techno-society question neither the values nor goals of their society. Like all other dominant cultures that preceded them, they assume the superiority of their belief system and seek to "enlighten" the world. As with Conquistadors, Da'esh or the Roman Empire, they neither question their actions or the consequences thereof. Their actions are driven by an unquestioning belief in the superiority of their cultural values. There is no one more dangerous than someone convinced that he/she is right.

Techno-culture, however, is different from many other cultural systems in one significant way. It matters little what the cultural beliefs are of the Afar hunter-gatherers or of herdsmen on the Mongolian steppe in terms of the effect upon the rest of the world. But the effect of the imposed techno-culture has consequences for every individual on the planet whether they want to participate in it or not.

It is as if the potential for everyone to be equipped with a smart phone is somehow the answer to the world's problems. Having access to all the information and technology in the world is useless unless one knows how to use it. A smart phone is only as smart as its user. By extension, technology works only as well as its operator. Potential does not equal outcome. At times the unexpected consequences of well-meaning actions are disastrous. Technological solutions often produce results aside from those desired or expected. The climate crisis is the most obvious manifestation of the unintended consequences of technological development. At the beginning of the machine age no one foresaw the impending disaster that humanity now faces. Likewise, the belief in technology to solve all problems prevents the necessary actions from even being discussed much less implemented.

Rather than take efforts to limit carbon emissions or protect biodiversity, the emphasis is on the development of more technology so that humans may continue to live without making changes. This unquestioned belief in technology as an agent of progress has created the nuclear nightmare and numerous problems from resource depletion to toxic waste disposal. Choking on plastic, humans have elevated technological development to the role of mythical savior.

Technology now reaches beyond the mere mortals who created it. The techno-culture is beyond the ability of nation-states and multinational corporations to contain it. Technology has become a way of life. It is seen as the solution to all problems, the ultimate enhancement. Technology has become the new religion for the modern world. Through technology all things are possible. It is the way, the truth, and the light-emitting diode.

1

The Deification of Technology

Any sufficiently advanced technology is indistinguishable from magic.
—Arthur C. Clarke (*Profiles of the Future, 1962*)

Any technology that can be distinguished from magic is insufficiently advanced. —Barry Gehm (*Analog* 1991; quoted by Gregory Benford, *Foundation's Fear,* 1997)

A quote widely attributed to Freud is "if god did not exist, man would have to invent him." In its original context, Freud, an atheist, was attempting to explain the attraction of religion. Suggesting that there is a psychological need for protection from the dangers of the world by a great all-protective force, Freud essentially hypothesized that human deities arise from psychological needs to feel secure.

The world has become a giant interconnected sphere with a host of problems from environmental degradation to a constant stream of terrorist threats. Rationally, there must be an awareness of the unpredictable and uncontrollable nature of events in a world with so many threats and competing ideologies.

The awareness of competing ideologies, whether religious or political, renders any notions of collective agreement nearly unattainable. Only one force in the current era appears to reach across cultures and nations. That force, technology, has emerged as dominant among all others. Its cultural superiority has become entangled with notions of good and evil. The temple of technology is all around. Devotees are always in touch with the godhead. The great imaginary machine in the cloud, through which all progress flows, is everywhere.

Many of the most ardent admirers of technology, especially digital technology, will likely be offended to be branded as a mindless follower of a cult that has made a deity of "0's" and "1's." But if one is unable to acknowledge the downside of technology, or exhibits beliefs that it can solve all problems in spite of evidence to the contrary, and that this ability is

5

absolute, then we are no longer in the realm of reason, we are in the world of belief and faith.

The contention herein is that the unquestioning and uncritical devotion to and acceptance of technology has become an impediment to a rational examination of the impact and effects of said technology. Due to beliefs such as the myth of progress and technological "solutionism" (Morozov, 2013), the embodiment of these ideas, technological innovation, has been deified.

The new religion for the modern age is technology.

Technology as deity serves to inflate the sense of importance for the human controllers. The development of advanced technology restores a sense of pre-eminence that has diminished in the human psyche over the past several hundred years as knowledge of the nature of the universe has slowly changed. When Copernicus put forth the Heliocentric Theory of the Universe he was widely challenged. He was not challenged because his logic was unclear but because he broke with religious orthodoxy. When Galileo later supplied the evidence by pointing a telescope at the stars, dogma refused to accept, or even examine, the evidence.

The stated theological reasons at the time were that these findings disputed the Bible and Christian doctrine. The actual reasons likely had far more to do with the diminution of the role of humans who could no longer view themselves as the purpose of existence. No longer being at the center of the universe questioned the centrality of humans in what was viewed as the divine plan. The associated knowledge that the universe was incomprehensibly large led to a severe psychological re-evaluation of the role of humanity.

The Theory of Evolution faced a similar rejection when first introduced. Theology argued that it refuted dogma. Actual reasons were again more likely related to a further reduction in the assumed special nature of humans. Copernicus and Galileo served to take earth and its inhabitants out of the center of the universe. Darwin served to remove the special role humans had reserved for themselves among all creatures (great and small).

Concurrently, and stretching across a time frame that encompassed both the introduction of the Heliocentric Theory of the Universe and the Theory of Evolution, there occurred the Industrial Revolution and the Machine Age. These periods of invention occurred at a time when discovery and knowledge led to questions about the place of humans in the cosmos. Control of machines and technology served to reinstate a sense of mastery and superiority.

The ultimate expression of these, splitting of the atom and putting a man on the moon, were achievements that restored a sense of superiority to humanity. Through it all, assisting in the quest, even guiding the way, was technology. With both endeavors the technology needed to make them happen had to be developed from scratch. Building these awe-inspiring machines psychologically elevated humans. No longer the purpose of the universe, humans created machines that controlled and explored the world. The death of those who died in service to building the bomb and reaching the moon were hailed as necessary for a higher purpose.

Cognitive bias leads to a tendency to remember the cases that support the argument, while disregarding those that refute it. The space program allows humans "to boldly go," and is heralded as an indication of how far we have come while the proliferation of weapons of destruction is shrugged off as a minor issue. Why worry about nuclear cataclysm when there is a new must-have app?

The disciples of the new religion are certain of the superiority of their beliefs and are dedicated to getting others to ascribe to their faith. Like Christian missionaries during the Crusades, conversion is a necessity and resistance must be eliminated so that all can be free. In every religion the fanatics attempt to force everyone to follow their belief system, willingly or coerced. The disciples and evangelists, the Crusaders and the jihadists, easily convince themselves that all of their actions are divinely inspired.

Like religious belief, technological deification serves to ignore the problems of the here and now with the promise of a better future. A final reward, the big payoff—these are the promises of a techno future without toil or problems. A quote from the cover of *Time* magazine concerning quantum computing serves to exemplify the implicit belief system surrounding technology—"It promises to solve some of humanity's most complex problems. It's backed by Jeff Bezos, NASA, and the CIA. Each one costs $10,000,000 and operates at 459 degrees below zero. And nobody knows how it actually works. The Infinity Machine" (17 February 2014). Solving problems without understanding: that is as close to a miracle as it gets.

The deification of technology along with the accompanying myth of progress are leading to massive congregations of unthinking, unquestioning adherents blindly accepting whatever dictates are pronounced from on high. There are occasional doctrinaire squabbles, Apple vs. Microsoft, Google vs. all other search engines, but the belief in the deity never wanes. Technology will save the world is the hype and there is no need to make any changes except upgrades.

The thesis that technology has become the religion of modern life might be viewed as hyperbole. The claim is not put forth lightly or frivolously. It must be explicitly stated that the deification of technology, i.e., technology as religion, is not simply a metaphor for explanatory purposes but rather is a direct statement that technology and its accompanying dogmas of progress and solutionism meet the standards for classification as a religion.

In every way imaginable, from the organizing and promotion of a belief system with a reliance on dogma, to the existence of disciples and evangelists, extending even to a promise of eternal life, the Church of Technology contains the structure and actions of an organized religion. A belief in technological progress as the means to human advancement is not questioned. This belief motivates the disciples to proselytize and FORCE, for the good of the forced, to convert to the "true" path. To do otherwise is to be in league with Satan (aka Dr. Ludd). Like the Crusaders of old, today's religious warriors spread their culture with delusions of "saving" everyone. Computers to the rainforest are just the modern equivalent of bibles in Latin.

Religion Defined

Consider the following definitions of religion:

"Religion is the 'feeling of dependence' of man on a higher power." —Schleiermacher. "Religion is what a person does with his solitariness." —Alfred North Whitehead. "Religion is the human relationship to a realm that is invisible and powerful and good, these three quantities combined." —Huston Smith. "A developed religion is an integrated system of beliefs, lifestyles, ritual activities, and social institution by which individuals give meaning to (or find meaning in) their lives by orienting themselves to what they take to be holy; sacred, or of the highest value."—Julia Corbett. "Religion: voluntary subjection of oneself."—*The Catholic Encyclopedia*, 1913.

The last quote, from the *Catholic Encyclopedia* might be the most illustrative and it also serves to demonstrate a perpetual blind spot of religion, which is to see only the positive. To be accurate, religion is also involuntary subjugation, maybe not to a god, but to the religion. Further examination reveals a goodness of fit between technology and its promotion to deity.

Digital Age Evangelists

One does not have to look far to find the champions of the digital age preaching their message of the salvation through technology. In *The New Digital Age* (2013), authors Eric Schmidt, former chairman of Google and Jared Cohen, Director of Google Ideas, put forth a not surprising utopian vision of the future marked by better living through technology. An overwhelmingly glowing take on technology; the authors unquestioningly and likely unconsciously subscribe to the philosophy of solutionism. Their approach to all topics is that technology will eventually eliminate all problems through slow incremental progress. Brynjolffson and McAfee, in *The Second Machine Age* (2014), likewise state "the first machine age was about overcoming the limits of physical being. The second machine age is about overcoming the limitations of our own minds." In many ways their views differ little from what Bernard Gendron (1977) long ago called the Utopian View of technology. Gendron, however, was mostly speaking in a theoretical sense and knew it, while the current harbingers of this message are true believers seeking to live out their faith. It is the promise of everything for nothing.

In an interview Schmidt stated unambiguously that "technology will reduce terrorism" and will "make life fundamentally better." Cohen seemed to go even further stating that "we must ensure the rest of the world comes along" (PBS NewsHour, May 2, 2013). Both statements speak volumes about a philosophy that perceives itself as superior. Questions regarding the role of technology in aiding and assisting terrorism, or concerns about privacy and totalitarian oppressions were shrugged off by the authors as minor issues. Both appeared to minimize or disregard information that suggested technology might not hold all the answers or even that there might be unintended consequences to seemingly positive advances.

Despite these rather significant concerns and the obvious problems associated with attempting to solve a problem through the very means that created it, the evangelists of the Church of Technology will give no quarter. Another case of the proselytizing devotee is the case of the apostle, Mark Z. Zuckerberg, of Facebook fame, has, as noted, developed a plan to wire the world. It is the building of missions all over again. The financial rewards that he and the other leaders of the companies carrying out their crusade stand to reap are, of course, incidental.

Eternal Life: The Goal of All Religions

A common thread among many of the world's religions is the belief in an afterlife. For some religions, especially Christianity, eternal life in heaven is promised to those who live according to the tenets of the faith. This idea, eternal life, is a prominent feature of belief that assures a future orientation and sacrifice in the moment.

Many futurists are also drawn to the notion of eternal life. Technological advance has become the means to achieve this goal. As organic cells appear to have a limit on their lifespan (Hayflick, 1965; Witkowski, 1985; Ruiz-Torres & Beier, 2005), many futurists have gravitated towards machines/cybernetics as the answer. Transhumanism, or Extropianism, is one of the more esoteric approaches to this issue with stated goals of merging with machines. The essential question, however, is not whether this is achievable, but whether it is desirable.

Let us briefly consider first whether it is even possible. Foremost among these problems is the very issue of how the mind is conceptualized.

There is a very real debate among neuroscientists as to exactly how the brain is organized and how consciousness develops and is defined. At present computer models are used but few neuroscientists or psychologists would see a direct parallel. In many ways it sells the human brain short to compare it to a computer. Even if the brain functioned in a completely parallel manner it is still in the realm of delusion or fantasy to suggest that the ability to transfer consciousness is a remote possibility. At present this is an insurmountable challenge. Transferring to a machine is a greater hurdle than building a machine that might develop consciousness. In effect, given the current state of knowledge it is more likely that artificial intelligence (AI) will be realized than it is probable that human consciousness will be uploaded into a machine.

Transhumanism buys into the myth that infallible systems can be created and that through technology there can be an end run around nature. What transhumanists do not address is that everything has limits, and everything breaks. At present, machines are flawed and imperfect, just as are humans. To live forever, the machine must live forever—a constant process of transfer with nothing ever breaking down or going wrong.

Table 1.1. An Exhaustive List
of All Machines That Will Never Break

1.	4.	7.	10.	13.	16.
2.	5.	8.	11.	14.	17.
3.	6.	9.	12.	15.	18.

Beyond the problems of things breaking, there is a larger problem for the extropian vision to be realized. Unlimited and exponential growth in terms of computing power seems to be a necessary achievement to actualize the vision of the singularity in which humans and machines eventually meld into one. These positions are clearly based on a misunderstanding and misapplication of Moore's Law which has become one of the more entrenched bits of dogma for the new religion of technology. The acceptance of unlimited growth into perpetuity belies a poor understanding of elementary math concepts such as the approach to limits and ogival functions. Again, it is very much a belief driven system when the evidence is against a position that continues to be defended.

Technological and scientific issues aside, what of the desire for the singularity, when humans and machines merge? What if it could be achieved? Does everyone get to live forever? It is likely that such technology would be enormously costly, especially at first. Would the technology be controlled by the state or by corporations? Is the technology only for those who can afford it or does insurance cover it?

Speaking of insurance, what if everyone lived forever? What would be the costs? Would there come a time when cost rather than health forces you to turn off the machine?

What of human population? Would it explode out of control? Would there be controls on reproduction as a result of too many people? Would means other than natural death be used to reduce population? If so, what will be the means and who gets to make the decisions?

What of boredom? What would the effect of living forever, dependent on machinery, do to humanity's conception of itself? Finding out that the earth went around the sun caused upheaval around the world; what would finding out that our consciousness can just be plugged into a machine do to the collective psyche? The ability to create technology has far outpaced the ability to consider its implications. The ethical implications of innovation are almost never considered. When the collective meme is that technology has the properties of salvation there is little chance that the acolytes will question its direction.

Technology is neither good nor bad, but to ignore its potential to create problems rather than solve them is demagoguery. Social consequences, including the effect upon the larger culture, need to be matters of discussion before the fact, not after. The desire to "enhance" humanity appears to be but another example of how ethics can be completely overlooked or obviated by the desire for "progress."

Merger with the Godhead: Becoming One with the Source Code

In many religious contexts, followers attempt to merge with the godhead through such means as meditation, taking hallucinogens, prayer, etc. As noted in the discussion of efforts to use technology to extend human life, perhaps indefinitely, the "need" to meld humans with computers is seen as paramount. In a similar vein, the Terasem Foundation, is actively seeking a means to upload the human mind into a robot (Terasem Foundation Website, 2015). Couple the placement of a human mind into a robot with a brain that has been "enhanced" through technology and a being that is very much machine and very little human is produced. This effort goes beyond creating life; it is an attempt to evolve humans into deities, or at the very least, demi-gods. Merger with machines is seen as conferring the power to become "more than human." This idea that merging with machines will make humans superior, or even the use of phrases like "transcendent" to describe the human-machine merger, reveal an image of humans as inferior. It is this philosophical divide, concerning whether the machines or the humans are superior, that drives much of the debate about whether such matters as uploading consciousness into a machine are acceptable. Depending on where one falls on this question—the superiority of humans or machines—determines a great deal about one's attitudes towards this approach to the extension of life. While this possibility does not appear within reach in the foreseeable future, and there are several obvious hurdles in the path to such an achievement, the remote chances that this could be realized raise severe ethical issues.

The history of megalomaniacal individuals who decided they were gods (Egyptians, Aztecs, etc.) does not portend well for a future populated by the "enhanced." If there is any truth to the adage that "Absolute power corrupts absolutely," the proliferation of a class of enhanced individuals is almost certain to produce at least a small percentage who will consider themselves god-like, if not actual gods. The unintended consequences of these actions must not be taken lightly. Therein lurks the danger in the technology-as-god approach that has subsumed the modern world. Uncritical acceptance of all that is labeled as advanced technology is not the mark of a rational being; it is the sign of a religious fanatic.

The Effects of Technology as God

The effect of engaging in the deification of technology is the creation of a dogma, as in all religions, and an unquestioning acceptance of all things associated with the deity as good, even perfect. There are several direct consequences that result from this belief system. Each of these poses some degree of risk for non-believers and humanity at large. Some of the primary effects as regards acceptance of the dogma underpinning a religion of technology are:

> Future orientation that sacrifices the present
> Confabulation between technology and science
> Belief in infallible technology
> Alienation from nature

The above list is by no means complete or exhaustive. However, these features have significant consequences and will be considered in turn.

(1) Future Orientation That Sacrifices the Present

The myth of progress and the concomitant belief in the ability of technology to solve all problems tends to be perpetuated by an ever-extending future orientation. No matter how poorly technology may perform and no matter how negative the consequences, there is always the promise that it will run/work/perform flawlessly at some point in the future. That future never seems to arrive and is always at some distant point. A perusal of content in various publications serves to illustrate this point. This is not just a reference to online media but also to the whole of the publishing world. Even in something as staid as print journalism this effect is easy to see. A sample of *Scientific American,* a magazine aimed at an educated market that attempts to present science in a clear and accurate way, over the past 20 years serves to demonstrate this effect.

In 1996 an article appeared touting "smart rooms." Technology that is still in the budding stages was hailed as inevitable and impending more than 20 years ago (Pentland, 1996). Five years later the same publication heralded advancements in what was then called the "Semantic Web," which would now be seen as fanciful if not ridiculous. While some of the ideas in the article have emerged (e.g., improvement in search engines and some of the Big Data confluences thereof) most of the ideas presently fall into the realm of fantasy, particularly the contention that it would "assist in the evolution of human knowledge as a whole" (Berners-Lee, Hendler & Lassila, 2001). In 2007, Bill Gates wrote an article predicting smart devices would control all the functions of a home in a time frame of less than two years (Gates, 2007). Basically, a restating and reinvention of concepts put forth in the smart rooms article by Pentland from a decade before, these concepts are somewhat closer but still not common. Note that two of these articles were penned by two of the "prophets," Bill Gates and Tim Berners-Lee. Only an apostate or heretic would dare question words from on high.

The flaws in smart technology become more apparent as the technology advances (see "Smart Everything Is Dumb," Reuters News Service, 2014). When initially proposed there was almost no discussion of any problems that might arise. The overly optimistic time frame regarding achievement of these ideas is typical of the introduction of new technology. What is introduced as flawless revolutionary technology turns out to be all smoke and mirrors. It also turns out to be a hacker's preferred mode of entry. Garage doors, thermostats and other common items connected for online control are now used to digitally break into houses.

Taking the perspective that it will eventually work in a flawless manner is a little like meditating to achieve nirvana. The possibility of the future perfection allows the adherents to overlook the discomfort and irritation of living in a constant Beta-test.

(2) Confabulation between Technology and Science

One of the great confabulations in all of history is the equating of technology to science. The two terms are often used interchangeably despite abundant evidence to the contrary and clear semantic distinctions. Technology is defined as:

> **tech·nol·o·gy** ... noun, plural technologies 1. the branch of knowledge that deals with the creation and use of technical means and their interrelation with life, society, and the environment, drawing upon such subjects as industrial arts, engineering, applied science, and pure science. *http://dictionary.reference.com/browse/technology*

At its core technology is concerned with the application of various fields of knowledge and at a practical level is largely about the production of machines and other devices. Science, on the other hand, is defined quite differently:

> **sci·ence** ... noun: 1. knowledge about or study of the natural world based on facts learned through experiments and observation: a particular area of scientific study (such as biology, physics, or chemistry): 2. a particular branch of science: 3. a subject that is formally studied in a college, university, etc. *http://www.merriam-webster.com/ dictionary/science*

Science, in contrast to technology, is primarily a quest for understanding and the examination of various ideas. In *The Structure of Scientific Revolutions* (1962), Thomas Kuhn notes that rather than a steady march to the truth, science progresses in fits and starts. There are many blind alleys of research, many ideas that fail to pan out. Fields of study can appear completely static for decades and then experience a frenzy of development. Change is often abrupt and unexpected yet follows a predictable pattern. In science, radical change occurs only after a preponderance of evidence collapses the old theories. When a field of knowledge undergoes a paradigm shift the adherents of the old paradigm are metaphorically overthrown. Those who embrace the new paradigm move from outsider challenging the status quo to the establishment figures. The new paradigm then becomes the "truth" until replaced by another viewpoint.

Science moves by paradigms and theories. Theories are slow to fall and slow to be accepted. Evidence guides the process. It can be argued that one of the positive checks in scientific process is the unusually slow pace of the peer review process. Poorly conceived and unproductive lines of inquiry are discarded in a manner that saves time and resources. In science, doubt is the key to knowledge, and nothing is ever settled. In religion, it is the reverse. Doubt is seen as a weakness, and everything has long since been settled.

Not only do paradigm shifts occur in science, they also occur in culture. However, cultural shifts often occur for reasons that have nothing to do with evidence or thoughtful analysis. The present time appears to be such a moment in history. Society is beginning to experience a cultural shift as a result of rapid technological change. The current truth, set forth by the current cultural paradigm shift, is a future of better living through technology.

The history of science again provides a useful lens of contrast. Science is replete with ideas that eventually did not pan out or were in turn supplanted. In science and academia, such failures may have no real effect outside the careers of a few individuals. Cultural shifts, on the other hand, have had disastrous results when viewed through the long lens of history. The food economy of Ireland shifted almost entirely to an imported, non-indigenous crop, the potato, which in turn had direct implications on the health and migration patterns of the population. Likewise, the introduction of gunpowder and other weapons in Europe that coincided with colonial expansion produced devastation and subjugation of numerous other cultures.

A more recent example of this might be the shift from a largely small business and subsistence fishing economy to one of international finance that occurred in Iceland in the first decade of the present century. Easy money and laws that brought massive foreign investment caused the economy of the entire nation to shift. The change was irresistible from a monetary standpoint right up until the whole house of cards collapsed with the advent of the worldwide economic recession. In retrospect, the whole of the banking system seemed to be operating on faith rather than sound business principles. Might the same phenomenon be occurring with the technological revolution?

Which orientation do the adherents of technology more accurately mirror, science or religion? Is there a paradigm or guiding theory to technological development? Or is it more "full steam ahead"? No time for questions. Technology is motivated by expanding its reach, finding new recruits, and creating more. Those who question aspects of technological development are often labeled as heretics or insulted as Luddites. Occasionally, technologists will use the label "un-scientific" to challenge critics.

While it is true that scientific principles are often used in the process of developing technology this is not a requirement. The history of invention is replete with those who had little understanding of the scientific process. Thomas Edison, progenitor of many of the devices of the modern age, relied mostly upon trial and error. Similarly, many of the great innovators of the digital era, such as Steve Jobs and Bill Gates, along with many others, were self-taught college dropouts with very narrow backgrounds outside the area of information technology.

This leads to an unfortunate state of affairs as regards inventions and their implications. The problem lies in ethics. Scientists are usually, at a very minimum, trained in the precautionary principle, which suggests the need for caution in considering the implications of one's actions and discoveries. In the field of psychology, ethical issues are more prominent because of the use of humans as experimental subjects. There have long been concerns about the implications of research into such areas as mind control, resulting in a tacit agreement not to conduct research into these topics. The human experimentation conducted by the Nazis in World War II is considered so out of ethical bounds that it is considered improper even to examine the findings.

There are unethical scientists, and there are certainly scientific advances that occurred because the unfortunate implications could not be predicted at the time, but the point is that science does directly and explicitly examine the possible consequences of discovery. The precautionary principle dictates that questions concerning the possible implications of actions be considered before the fact. A direct manifestation of this can be seen in the requirement that any research involving human subjects be examined by an outside group of reviewers to determine if there are any risks to the people involved. Institutional review boards are required whenever any research that involves humans is going to be conducted. Couple this requirement with the precautionary principle and the first question in science becomes "Should this be done?"

Technology, by contrast, asks the question, "Can this be done?" Technology is interested in how things will be done and how they will be applied but is woefully deficient in terms of ethical analysis. In many ways all that limits technological experimentation is what material it is possible to obtain. Certainly, some of this relates to the profit motive and consumerism. However, much of this lack of attention to ethical issues is a product of the inability to consider the implications amid the rapidly changing nature of the technological landscape. Everything has become transient and there is no time to think about that which will be obsolete tomorrow. This might not be so important if technology did not possess the potential for such devastating outcomes.

A literary example draws out this idea. In Kurt Vonnegut's novel *Cat's Cradle* (1963), the world is brought to an end by a scientist who made an isotope of water that freezes at 100 degrees Fahrenheit. It seemed like a good idea until all hell broke loose. And so it goes with much of modern technology.

Thinking of ideas is not the same as thinking them through. Not everyone is a great philosopher. A Hitler comes along about as often as an Aristotle. Adherence to a precautionary principle is likely more important now than in the past. Previously, cultural lag served as something of a check on poorly thought out innovations and persons with nefarious purposes. Lack of connectivity between peoples also served to limit the danger a bad idea or invention could have. Outbreaks of diseases tended to be geographically contained because of the inability of most individuals to travel very far from home. Now because of air travel, a case of measles at Disneyland spread across the nation and the whole planet panics due to an Ebola outbreak in Africa. The Covid-19 pandemic spread throughout the world before it was detected. The initial response has been the decidedly "no-tech" approach of social distancing. It is noteworthy that pollution levels dropped significantly in many places in the first few weeks of the lockdowns around the world.

In the present era, technological innovation proceeds so rapidly that the pace tends to obviate ethical analysis. This leads to a blind spot among inventors as regards the possible consequences of innovations and breakthroughs. The race is to market, not to the philosopher.

Many brilliant minds give little or no thought to ethics. Especially in fields such as the physical sciences the world is often viewed as a riddle to solve. Interacting with chemicals, math, and the physical world is viewed as involving only intellect and logic. Ethics are viewed as the domain of philosophers and social scientists. This has the consequence of there being something of a lack of ethical analysis on the part of inventors as regards the possible misuse of the results of their work and even less thought given to the unintended consequences of technological breakthroughs.

Rather than tech developers, much of the philosophical considerations of technology have been made by authors of fiction and science fiction, such as Vonnegut, Ray Bradbury, and Philip K. Dick, rather than by scientists.

A notable figure and exception in this regard would be Isaac Asimov, who was not only a science fiction writer but also a scientist. Asimov, known for his *I, Robot* trilogy (1950), has come to dominate and permeate modern culture as regards the ethical considerations of technology. Asimov is also the author of one of the primary bits of dogma in the Church of Technology. He put forth the so-called "Three Laws of Robotics" in his early work (Asimov, 1942). The "Three Laws" are as follows:

A robot may not injure a human being or, through inaction, allow a human being to come to harm.

A robot must obey the orders given to it by human beings, except where such orders would conflict with the First Law.

A robot must protect its own existence as long as such protection does not conflict with the First or Second Laws.

While these "laws" figure prominently in science fiction and in the minds of those fascinated with robots, it would be incorrect and potentially fatal to assume that these laws are always followed and kept in mind. Many technophiles seem to feel that the "laws" are all that are necessary to insure a future nirvana. This also belies a belief that technology can be perfectly controlled. Again, what could go wrong?

Similar "laws" regarding the creation and use of nanotechnology have also been developed. It is important to note that while these ideas are framed as "laws" there is no real regulation of them or many other technological matters. This exists even though there is evidence that nanotech could pose some very real dangers in the organic world (Graham et al., 2017; Elsaesser & Howard, 2012; Oberdorster, Oberdorster & Oberdorster, 2005; Rivera Gil, Oberdorster, Elder, Puntes & Parak, 2010). A primary concern regarding nanotech is that it could "escape" the laboratory or other controlled environment. The consequences of this are not only unclear, they are difficult to estimate. The Center for Biological and Environmental Nanotechnology (CBEN) found nanoparticles accumulating in the bodies of research animals as far back as 2002 when the technology was relatively new.

In many areas, such as the efforts to attain eternal life through uploading consciousness into a machine, little consideration seems to be given to what might be the consequences of achieving this goal. Would everyone get to live forever or only a few select individuals? Should only some individuals get to live forever, what are the parameters for selection? What would be the effect on human population and resource depletion? Is working towards this goal, which many argue is completely implausible for a wide variety of factors, an efficient or ethical use of resources? Even questioning the validity of such efforts is an affront to those who see only the promise of belief.

(3) Belief in Infallible Technology

Despite numerous failures, and a host of problems created by technological innovation, the belief in the ultimate infallibility of technological advance reigns supreme. Inherent in the development of many technologies is a belief that it is possible to design a fail-safe, flawless piece of equipment.

The problem with the belief in infallible technology is that it leads to the creation of disastrous scenarios that would never have occurred had there simply been an understanding that machines break, plans fail, and nothing is foolproof. The limits of technology are constantly exposed with the disappearance of the Malaysian airline flight MH-370 serving as a stark example. The plane, a Boeing 747, disappeared without a trace. At every stage of the search, technology was introduced that was supposed to answer the question of what happened. Multiple means of tracking the plane yielded no results. It was only when parts of the plane washed up on shore was there any idea what had happened.

Tracking a moving object is a relatively well developed and understandable technology that has existed for quite some time yet is still far from perfect. If it is not possible to find a 747 airplane, equipped with tracking devices, how advanced and perfect should technology be assumed to be?

Many religions have a dictum to the effect that "One cannot understand the mind of god." Technological deification reflects this in a preference for increasingly complex devices. In fact, technology that is incomprehensible seems to a cultural preference. By and large, the understanding of technology is far more limited than the use. The need to always be on the cutting edge has led to an obsession with gadgets and a preference for the overcomplicated, and at times, risky.

Technology properly defined makes little distinction between machines and devices that are complex and those that are simple. Fire is a technology; a lever is a technology. Many of the most important technological breakthroughs in the history of humanity were simple devices that are still used as originally developed. However, in the mindset of most people, technology has come to mean a complicated machine. This misconstruction is directly related to the deification of technology.

The very belief that technology can be infallible leads to an insidious outcome. This outcome relates to an overt preference for complex technology over simple technology. The more complex the better it appears. The increasing complexity of gadgets and devices psychologically plays into the image humans have of themselves as brilliant and above the rest of the animal kingdom. Examples of this approach include a preference for nuclear energy over solar power and cars that drive themselves for safety rather than the proven strategy of putting roll cages into automobiles.

Any technology, even those that have obvious and unmitigated benefit, can be misused. An example from a different realm of technology demonstrates this with alarming clarity.

Additive printing, also known as 3-D printing, is a clear and unmistakable advance in many ways. One of its major applications has become a crowd sourced method that has allowed for the design and manufacture of prosthetic limbs in a manner that dramatically reduces the costs to prices below $100 (PBS News Hour, January 16, 2015). It is difficult not to see the social good in the production of prosthetics for poor children. Yet, while this technology can be used for an unmitigated good, it also has the capacity to completely destroy the manufacturing sector as it is now understood. More alarmingly, there has already been a demonstration 3-D printing production of a gun that was able to fire bullets. This problem has moved from theoretical to reality and the complication it causes for matters such as airport security and gun control efforts is incalculable.

Once again we arrive at a point where a technology has been created before we are able to determine the possible consequences. Developments such as AI and nanotechnology lead to the issue of creating technology that must be flawlessly controlled and maintained. When one small mistake can lead to a catastrophe—when there is no room for error—all that can be assured is that eventually something will go wrong.

One of the poorly confronted problems with technological development lies with the very human factors of intelligence and personality. Intelligence is not related to personality in any straightforward manner and it is unwise and unsupportable to assume that those with the intelligence to create a technology are also those who will be able to consider the possible misapplication of their inventions. Many scientists, like people in many fields, are naïve and somewhat one-dimensional in their personalities. Dedication to pure logic and rationality may make many scientists oblivious to the possible consequences of their actions. The mental torture and anguish that many of the scientists involved in the Manhattan Project later professed stands as testimonial to this idea. Robert Oppenheimer's words at the detonation of the first atomic bomb, "My god, what have I done," along with his "I have become Death" speech a short time later, stand in stark confirmation of this truth. Einstein was known to have been haunted by his part in the creation of the bomb, wondering whether any of it could have been done had he not developed the Theory of Relativity.

As noted, when psychologists conduct experiments they are required to submit any proposal that might involve human subjects for review by a professional board that examines the study for any possible harm that might come to the subjects (APA Code of Ethics, 2015; United States Public Health Service, 1966). Those in fields such as robotics, AI, mechanical engineering, etc., however, face no such constraints other than those that might exist on any materials, such as chemicals, they might seek. The out, as it were, is that they are not working on humans. However, the potential effects on humans might be enormous in some cases. While a psychologist might harm one individual, some technology has the possibility to end everything.

If the precautionary principle were even minimally acknowledged there would be many technological advances that would never have taken place. However, as with the case of the nuclear bomb, it is easy to willfully disregard the possibilities or rationalize away the danger because of circumstances. Gene editing technology is an emergent area where ethical considerations are minimized as a result of the possible positive effects. Understanding the effects of altering our evolution seems secondary, at best.

Even if somehow, every piece of technology developed was fail-safe, if it could be completely controlled and maintained without threat (like we are supposed to believe the nuclear arsenal is), there is still the problem of the human operators, or more specifically the philosophical problem of the bad actor. A human redirecting, misdirecting, misusing, take your choice, certain technologies can have tragic consequences for the entirety of humanity and the planet. It would be nice to believe that the idea that there are people who want to blow up the planet or cause some sort of plague is just a science fiction nightmare.

It would be nice, but it is not the truth. At this very moment, there is the very real concern that a terrorist group could assemble a dirty bomb. Biological agents can be easily created with only minimal knowledge. More than three decades ago followers of the Bhagwan Shree Rajneesh poisoned a salad bar in Oregon (Fitzgerald, 1986).

It doesn't even have to be intentional. The panic surrounding the Ebola outbreak of 2014–2015 attests to the capacity of air travel to aid the spread of disease and irrational behavior. How long before some terrorist group decides to deliberately infect some of its members with something more easily spread than Ebola with symptoms that remain hidden until after it is too late? One individual now has the capacity to wreak havoc upon everyone.

Let us not forget political ineptitude. Look at the history of regulation of the global arms trade and weapons in general. Only in April of 2013 did the United Nations finally pass a treaty regulating the international trade in weapons (Crane, 2013). Firearms, especially in the United States, continue to be more loosely controlled than permits for garage sales.

The total lack of coordinated government response to climate change, or even minimal regulation of chemicals in the environment, often in the name of the free market, all belie the dangers of unrestricted and unregulated technological development. When a significant number of the political figures making decisions that affect the whole world believe that any government regulation is a subversive plot, what chance is there for thoughtful discussion of the direction of technological development. Too little, too late, seems to be the sum total of regulation.

(4) Alienation from Nature

Belief in the supremacy of technology has the unintended consequence of reducing nature to a secondary role. Destruction of the natural world is rationalized. Mining and destruction of habitat are unquestioned as belief in the possibilities of new technologies makes it appear unnecessary to safeguard the natural world. Ideas of off-planet exploration and colonization are attractive options to doing the hard work necessary the clean

up the world. It is the miracle as solution. Rather than attempting to preserve the planet the response is to find another. This is analogous to putting
a down payment on an expensive beach house while your current home is
in foreclosure.

Another such example of the lengths to which technology is employed
in the demonstration of "everything is possible" is through efforts to revive
extinct species. This effort, which ignores the multitude of factors that led
to a species' becoming extinct, seems more about gathering attention than
producing anything of scientific value. Efforts are underway to resurrect
numerous species using technology ranging from selective breeding of
related species to attempts to clone mammoths and dinosaurs from recovered DNA (Zimmer, 2013). Jurassic Park aside, the effects of reviving species seems fraught with the possibility of unintended consequences. One
possible consequence of actions such as bringing back extinct species or
geo-engineering is that once humans can do these things, it makes it even
less likely that we will actually attempt to address the attitudes that led to
extinctions and climate change to begin with.

Technological fetishism is a view that technology has become the master and we the slave (Gendron, 1977). In Gendron's largely pro-technology
text he noted the dystopian view that "Technology is growing on its own,
not as a consequence of free human choice, and technology, in its growth,
is shaping all the major institutions and practices of society" (pp. 160–
161). He further pointed to what he called a "compulsive" use of technology. Whether it is through a compulsion or a sense of religious devotion, it
has become clear that the application of technology to all aspects of life is
pervasive.

It is this inability to examine the effects of technology in a dispassion-
ate manner that is preventing humans from acknowledging the source of
their problems, the very technology that is hailed as progress. Yes, technology is making some aspects of life easier and is leading to interesting collaborations and connections. Unfortunately, the costs, in terms of
resource depletion and the excessive and continually burgeoning demands
for energy, are unsustainable. Only a fanatical belief in divine intervention
could rationalize continuing down the current path. Yet, no matter what
the consequences, no matter what the negative effects that might be produced by a technology, all problems are rationalized away as steps on the
road to a better, more perfect future. Underlying all—supporting the ideas
that technological dependence is necessary, that technology is a mythical
savior, and that everything is on the right track—is the myth of progress.

2

The Myth of Progress

Is it progress if a cannibal uses a knife and fork?—Stanisław Lec

Every civilization that has ever existed has believed itself to be at the pinnacle of human achievement. This holds true even of societies that collapsed because of their own excesses or stupidity. Whether it was the ancient Romans spiking their wine with lead or the Aztecs engaging in human sacrifice, humans have consistently shown an unerring belief in their own superiority and an ability to ignore misguided actions.

As civilizations rise and fall there is the unmistakable and seemingly unavoidable tendency for each society to see itself as "better" than that which it replaced, superseded, or conquered. This type of superiority complex underpinned the actions of the Conquistadors, Moghuls and Nazis, to name only a few. Given the carnage that has resulted from just these few examples, how then can this view be justified?

Rather than a recognition that the place in life attained by the "glorious empire," or "royal family," or whatever other grandiose self-justification for having what others do not, is a byproduct of random events, the alternative hypothesis of divine influence or moral or intellectual superiority is selected. Uses of phrases such as "chosen people" or "the elect" serve to reinforce the idea that advantage and disadvantage are the result of the inherent superiority of those in positions of power.

The belief that the world is just creates a mindset that seeks to confirm that the differences between the "haves" and "have nots" as regards access to the assets of society is a result of flaws in those who are without. Belief in a just world, the Just World Phenomenon, is an experimentally validated cognitive construct (Lerner & Montada, 1998; Furnham, 2003) that serves to reinforce the idea that people get what they deserve. Belief systems, whether political or religious, seek confirmation in the differences between those who follow and prosper and those who deviate and suffer.

Differences become the focus for comparison. Judgment of these differences is known as cultural relativism. Every society has adopted certain beliefs and customs that it decides are the only "correct" way to do or think. A dominant culture tends to assume that because it has become dominant its views are superior and legitimate. In other words, might tends to be interpreted as right. While this view is widespread, there are few cultural absolutes. There are few matters of life that are absolutely followed by 100 percent of the cultures on the planet. People eat with forks in one place, chopsticks in another, and with their hands in a third place. Someone who is a cousin in one culture may be considered a brother in another. In some languages reading is right to left, in others, left to right.

Yet, differences are mostly in form rather than substance. No matter what one eats or how, eating still occurs. No matter the way kinship is reckoned, it is tracked in some manner. Whether one speaks tonally or not, language is an important component of what it means to be human. Yet most humans continue to think of their nation, culture, group, etc., as superior. This notion of superiority is the source of national and ethnic pride but also of jingoism and racism.

Cultural relativism is not only about humans comparing themselves to other groups living concurrently. Humans also tend to think of themselves as being superior to those in the past. By extension, humans of the future will be even greater. Rather than focusing on the fact that modern humans continue to live in a manner similar to that lived for millennia, in small social groups, struggling for survival and existence, with the need for food and shelter still preeminent, there is a tendency to concentrate on trivial differences. Again, the changes tend to be more form than substance. The ancient Egyptians may have written on papyrus as opposed to a word processor, but the important aspect of the activity was writing it down to preserve it.

In addition to cultural relativism, the idea of progress is also distorted by the relatively brief span of human life. A broad view of the history of a universe that spans approximately 14 billion years of existence is difficult for a creature that lives less than 100 years. Even though humans have been writing it down (the important part, as noted above) for several thousand years and as a result have access to a type of cumulative knowledge, the scope of any one person's experience can only be for a few generations. This leads to comparisons of the present to the very recent past. The selective nature of memory leads to remembrance of negative events and a focus on those things that changed, such as technology. The removal of negative features of the recent past and the changes that occur in the lifetime of one individual become linked in a manner that incorrectly implies cause and effect. These changes are then framed as progress.

It is true that technological advance has led to innumerable improvements in the quality of life for vast stretches of humanity. From the eradication of smallpox to the substantial decreases in infant mortality, life has been altered in a manner that can almost inarguably be viewed as a social good. However, at the same time these advances were occurring, human population has exploded (more than doubling in the last 50 years), and diseases like HIV and Ebola have manifested. We live longer but are less healthy. War is sanitized, with killing more efficient and the deaths at a distance. In a sense, all that has been done is to trade one set of problems for another.

The technology of every society/civilization inevitably becomes the focus of that culture. Just as the bow and arrow was the focus of hunter-gatherer societies, computers have become the focus of the digital society. Most people no longer hunt to get their food, but everyone still needs to eat. The bow and arrow are gone but the essentials of life remain the same. To eat and reproduce. The computer as we know it will also eventually be supplanted, but the essentials will remain the same. Technology is neither the purpose nor the endpoint, it is merely a tool. To view technology as an end in itself is a fundamental error. It is not the development and possession of technology that is important but rather its ultimate effects.

Technological innovation has produced machines that do much of the work previously done by humans. These same machines have eliminated countless jobs and produced enough carbon emissions to alter the climate. Volcanoes and natural disasters formerly provoked superstitious efforts to control the forces of nature. Today, technological promise provides an illusion of security that all will be well, while nuclear annihilation hangs over humanity like a modern sword of Damocles. Volcanoes, earthquakes and tsunamis are oblivious to technological innovation.

Many of the things which modern humans are so narcissistically impressed with being able to do so well are merely extensions of ideas conceived by the ancients. They created math, science, art, and philosophy. They built massive monuments like Stonehenge, the Pyramids, Gobekli Tepe, Chichen Itza, and the Sphinx, all without the aid of bulldozers or cranes, or any complex machines. Using simple technology and most importantly their brains, complex and amazing structures were built. It is not the technology that is truly impressive; it is the human brain that conceived it.

Despite the above, the myth of progress continues to be perpetuated. Every invention is heralded as a breakthrough and seems to prove that the steady march of technological innovation is never-ending. The societal assumption that progress is inevitable and linear serves as reassurance that all problems can be solved. This assumption, however, is more belief than reality. At the level of evidence, progress is more illusion than reality, more a matter of framing or perspective than actual change. Several examples of "advances" created by technological "breakthroughs" are illustrative of this point:

(a) The iPod by Apple was hailed as a major innovation upon release. The ability to have portable music was touted as revolutionary. It is the Walkman re-imagined with online capability.

(b) Mechanical keys have largely been replaced with versions that have computer chips. Thieves are still able to break into cars, they just need different tools. The real difference is in the costs of duplicating the keys which went from a few dollars to around a hundred dollars.

(c) Windows 8—Tami Reller, CFO of the Windows Division of Microsoft, said upon the introduction of Windows 8.1. "Windows 8.1 furthers the bold vision of Windows 8 by responding to customer feedback and adding new features and functionality that advance the touch experience and mobile computing's potential." Translation: Windows 8 was a dog and we're trying not to lose our customers.

(d) Galaxy Note 7—within weeks of its release pulled from the market due to exploding batteries.

(e) The Internet based economy now allows people to shop from the comfort of their home and have everything delivered. A hundred years ago most people ordered everything from catalogs and had it delivered to their door. Then there were very few stores, now the stores are all closing.

(f) Autonomous vehicles or "driverless car" technology is currently all the rage. The very term "driverless" creates something of a false conceptualization of what the technology is about. Driverless really refers to the idea that the passenger does not have to physically control the vehicle. The ability for a car to take a passenger where they want to go without actually driving is already well established. It goes by several names: taxis, limousines, buses. What driverless cars are really doing is using technology to supplant a human being.

Driverless cars also ignore the fact that many people find driving to be a pleasurable activity. Beyond the legal issues surely to arise in the case of accidents, the concept of a passenger sitting in an automobile while a computer drives is simply absurd at a basic level. Beyond any other consideration, putting more cars on the road, driverless or not, is **not** progress.

Efforts to improve automobiles over the past couple of decades have also demonstrated an increasing preference for digital versus mechanical technology. The automobile may be one of the clearest examples of a mechanical technology. Recent changes, however, have turned many of the formerly mechanical features of a car into electronic and digital control. From items as simple as windows and mirrors to the most complex braking mechanisms, functions that were formerly mechanical have now evolved. The result has been that cars have become increasingly difficult for individuals to repair without the assistance of a trained professional. Even trained mechanics now must plug the car into a diagnostic computer to determine what is wrong. The consumer who works on certain parts of the car voids the warranty. Minor functions controlled solely by digital chips now require expertise and expense to repair. As noted, making a copy of a key is no longer cheap but somewhat costly and for some cars, effortful.

Car ads routinely tout enhanced safety features that generally involve technological advances such as sensors that notify the driver of cars too close, anti-skid brakes, cars that have the ability to override the driver and take control. Yet the singular most effective means of protecting drivers from car crashes, roll cages, proven effective even at the high speeds of racing, are rarely placed in cars. Adding at most a couple of hundred dollars to the price of cars, a technology that would also have the effect of allowing cars to be made lighter and therefore more efficient, is almost nonexistent in practice. The fact that a roll cage must be installed at manufacture and cannot be up sold as a feature adds cost but not profit. Simply put, progress is often defined in terms of the technological innovation, not any necessary improvement in outcomes.

More is not always better and new is not always progress.

An intuitive understanding of this proposition can be attained by conducting a minor thought experiment in which a world without electricity or the burning of fossil fuels is imagined. For vast numbers of people what immediately comes to mind is a future where we are less harried but living in a more primitive state. Certainly, the view that most would have of pre–Industrial society is inaccurate but the projection of a life with less stress is not only common but generally accurate. Is the tradeoff worth the hassle?

Consider the period of history surrounding European colonial expansion that occurred in the 15th through 18th centuries. In almost every corner of the world life was arguably less stressful and subjectively better before the forces of imperialism brought alien technology, religion, and culture upon their societies. It was a technological advance in shipbuilding and navigation coupled with innovation in weaponry that allowed Europeans to essentially subjugate the rest of the planet. For the Europeans this worked out well but not so great for everyone else. Progress at times is a matter of perspective.

Philosophical musings aside, it appears difficult to not buy into the notion that technological innovation is a steady march of advancement and progress. Technology has become identified in the modern psyche with superiority. Rather than recognizing that the brain that is able to image, create, and innovate is the evolutionary advantage, many confound those items created by the brain with advantage. A better tool becomes an advantage. Psychologically, computers and other modern technology are analogous to a stronger atlatl or better method of flaking stone for paleo-humans. This tends to ignore those technological innovations that turned out to be worse than what they were to replace such as asbestos and lead paint.

History is replete with innovations that eventually proved unacceptable. The Hindenburg and airships of its kind are one example of this. The Edsel might be a more mundane case of reality not measuring up to hype. Quadraphonic sound anyone? New Coke? Progress is often defined by context rather than any meaningful change in effect. A milking machine might be progress for the farmer but not so much for the cow.

What survives is not necessarily that which is the most advanced or even the best when it comes to technology. Many of the decisions about which products and innovations are successful are not, as economists presuppose, the product of rational decision making after a process that invokes images of competing products being put to a contest with the best emerging the winner. Rather, the technology that advances and becomes dominant is the result of many complex factors that have little to nothing to do with quality, but more to do with matters such as who had the best marketing and PR. Consider the AC versus DC power debate that took place in the early 1900s between Thomas Edison and Nikola Tesla. Edison essentially won the day by manipulating the public debate. He electrocuted birds with Tesla's approach and forced us to plug everything into the wall. Today we are rediscovering Tesla and his techniques are used in Wi-Fi and other "new" technologies.

Contrasting the lives of Edison and Tesla is illustrative of the realities of how technological advance spreads. Nikola Tesla was one of the great intellects in history and demonstrated an understanding of the application of scientific principles far beyond that of almost anyone else who ever lived. He holds over 100 patents and figured out how to provide free electricity to the world. Edison, in contrast, had a poor understanding of scientific principles and relied almost exclusively on trial and error and dogged determination. Many have claimed that Edison invented very little but effectively exploited those working for him (including at one time Tesla) in order to patent ideas and benefit commercially (Oatmeal Website). Unlike Tesla, Edison was successful precisely because he saw commercial applications of ideas, not because of superior scientific knowledge. Tesla, on the other hand, was somewhat naïve and not particularly interested in money. Many of his ideas were originally credited to others who were better at marketing and more interested in money than science. A modern parallel, although less extreme, might be Steve Jobs and Bill Gates. Jobs appeared to have a vision of the products, while Gates had a vision of how to mass market.

This idea, that what survives or what dominates the market is necessarily the superior product or technology, is a type of flawed thinking Francis Bacon identified as the "Idol of the Marketplace" (Bacon, 1620). Although Bacon pointed out 400 years ago the fallacious nature of believing that which is popular is necessarily the best, humans continue to make this fundamental error in thinking. In Bacon's day individuals might be forgiven for such assumptions but a modern netizen who is constantly bombarded with efforts of psychological manipulation known as advertising cannot afford to be so uninformed to the effects of market forces on innovation.

In the documentary *Who Killed the Electric Car?* (2006) the decades ago development of the electric car is detailed. It is noted that the car did not fail due to problems in operation or due to a lack of demand. In fact, owners of the cars were fanatical and kept them running despite efforts to get them out of circulation. What killed the electric car was collusion by oil companies and other interested parties, primarily the dealerships. Solar power, wind power and other forms of alternative energy have effectively found the same barriers. A technology with great promise goes largely undeveloped and unpursued because it is not viewed as economically profitable. In the case of electric cars, the dealerships were particularly opposed to the cars because they did not need the type of servicing that brought profit to the dealership. The fact that the cars were relatively maintenance free worked against their acceptance. Betterment of the human condition is not even a close second to economic considerations as regards the promotion and development of technology. The information age is as much about propaganda and marketing as it is about information.

Producing useless material and wasting resources on the production of needless consumer goods is one outcome of this belief. The damage done to the planet is another concern. However, it is the long-term consequences of belief in unending progress that slowly warps the discussion in a less obvious way. Belief in the myth of never-ending progress creates a sense that the best is yet to come and that the past is just a burden to overcome. By extension, this myth also leads to the belief that any problem can be fixed, so there becomes little concern for the consequences of development.

The failures to address the environmental consequences of a technologically based society are connected to the belief that any problem can be fixed through innovation. Yet the Digital Age has not stopped the pollution of the Industrial Age, it has only shifted it. The shift is, perniciously, to a potentially more lethal type of pollution that, because it is out of sight, is out of awareness. Many people who readily accept human involvement in global warming argue against "reactionary" measures such as cutting fuel consumption. This group, highlighted by every State of the Union message for the past 30 years, discusses "new technologies" that will lead the nation into the future.

Relying on new technology to solve problems created by technology is a little like going on a spending spree because you have heard of a method to insure winning the lottery or beating the house in Vegas. When it comes to climate change, a more fitting analogy might be driving a car towards a cliff as fast as possible with the belief that a parachute can be built on the way down. Counting on a solution not yet developed in order to justify making a known problem worse is irresponsible and irrational: Technophiles minimize such a statement as reactionary. Never-ending progress is not only supportable according to this point of view it also has empirical support. The evidence that innovation can be indefinitely sustained is supposedly present in the construct known as Moore's Law. It is unfortunate for those making the argument for never-ending innovation that Moore's Law is the basis of their views, as Moore's Law makes none of the claims for which it is generally credited.

Moore's Law: The Misunderstanding and Misapplication of Ogival Functions

One of the more profound ideas that underpins the belief in a continual world of innovation and progress is a concept knows as Moore's Law. Originally put forth by Gordon Moore, one of the co-founders of Intel, the "law" is more accurately an observation. Moore's Law is usually reported as if it is an empirical fact and generally stated in such a manner as to claim that computer power and processing speed doubles every two years (often reported as eighteen months after a comment by Moore's Intel co-founder). However, while the common interpretation of Moore's Law concerns the doubling of computer power and processing speeds the actual statement upon which this is based is as follows: "The complexity for minimum component costs has increased at a rate of roughly a factor of two per year. Certainly, over the short term this rate can be expected to continue, if not to increase. Over the long term the rate of increase is a bit more uncertain" (Gordon Moore 1965).

Moore's initial comments were not even directly about processing speed or computer power but rather about circuit density and the ability to fit more integrated circuits onto a control board. It was not until 10 years later that Moore added his prediction about doubling every 24 months (Schaller, 1996). What is rarely reported or even acknowledged is that Moore himself predicted that this state of affairs "can't continue indefinitely," (Dubash, 2005) and went on to note that there are fundamental limits on how small things can get.

While Gordon Moore may recognize the limits of nature and the predictions of what has been termed his law, others clearly do not. It appears that a vast number of innovators, especially those in highly technical fields, including those with the mathematics background to know better, have accepted Moore's Law as an unassailable fact.

Even with the misapplication and misunderstanding of Moore's Law that is so common, there is no reason that computing power *must* keep doubling. To do so may be a giant waste of resources. Speed and capacity are already at limits that humans have a hard time comprehending. Humans can only consciously process somewhere between 50 and 200 bits per second (automatic human processing at the sensory level is different but not of concern for the current point). Most of the increases in speed of computing devices are imperceptible to the average person at this point. Increases in capacity are almost absurd in a very fundamental way. No one can fill up a hard drive in regular use. Yet the drives to increase speed and capacity are almost compulsive in the digital world. Efforts to sacrifice any of these features to develop a "green" computer, however, are nonexistent.

Since Moore made his famous statement, his "law" has been applied and misapplied to all manner of things including the biological origins of life (Sharov & Gordon, 2012), production of solar cells (Naam, 2011), automobile manufacturing (Benneton, 2008), data collection (Hadhazy, 2012; Rogers, 2011), and almost any area that even minimally involves technology of any kind.

These applications are simply off the mark. Moore was specifically talking about the number of transistors that could be produced and placed on a circuit board. It was an observation that related to the cost of producing his company's product and not intended as a general statement.

Despite the apparent widespread application and acceptance of Moore's Law there are very real reasons to question its over application. The first and foremost reason to question the manner in which it has been applied lies in basic math (detailed below) and the other reason is that the fundamental laws of physics place limits on many things. This appears to be Gordon Moore's own position based on an interview—"Moore's Law is dead, says Gordon Moore" (Dubash, 2005).

The inability to apply everything related to technology to Moore's Law is becoming more apparent. Even in the computer world, the law is beginning to break down. At a website entitled: "Moore's Law or How overall processing power for computers will double every two years" (www.mooreslaw.org) this idea is thoroughly debunked. It notes that speed of processing has barely doubled in the past 10 years even though the number of transistors has continued to follow the trajectory Moore outlined.

To rephrase, Moore's Law appears for now to hold up in terms of what it is actually about (number of processors) but does not hold up in application in the sense that there is not a one-to-one correspondence between number of transistors and computing power (www.mooreslaw.org).

When correctly applied and interpreted, Moore's Law is a restatement of a fundamental math proposition related to the cumulative curve. Assumptions related to continual exponential growth are often a failure to grasp a fundamental mathematical concept pertaining to cumulative effects and the limits thereof. It is this fundamental misunderstanding that allows for many of the previously noted applications of Moore's Law to such matters as the origins of life, automobile manufacturing, data collection, etc., to appear to have validity in their initial application.

Ogival Functions and the Cumulative Curve

In the field of psychology, especially that related to learning and the acquisition of learned behavior, it is common to graph the change observed in the acquisition of some type of knowledge or behavior. The number of learning trials or opportunities has been found to be directly related to the acquisition of the desired behavior in a predictable manner. The curve, or function, produced as a result of this graphing almost always appears as follows:

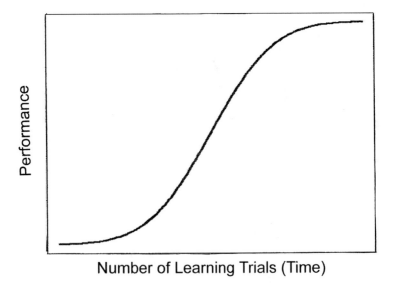

Figure 2.1 Standard Curve Produced in Learning Experiments

A way to conceptualize this curve is to think of it as produced by someone learning a sport. At first, the initial, flat part of the curve represents when the learner is attempting to understand the sport and acquire the basic skills. Performance is poor and improvement is slow. Then, as one "gets" the game, performance begins to improve rapidly. This is what happens at the inflection point where the curve suddenly starts to exhibit a severe slope upwards. At some point, the improvement in performance begins to diminish and even the most accomplished professional will eventually reach something of a plateau which is represented by the point at which the curve stats to flatten out once again. This is the point of diminishing returns. Gains may continue to be made but the individual essentially reaches a peak and future gains come at greater costs.

Physical limitations and other barriers serve to prevent performance from going beyond a certain limit. The records for various types of athletic performance demonstrate this effect as well. It does not seem to matter whether the sport is the high jump, the mile run or something else; the demonstrated effect is the same. At some point records in the sport experience a rapid improvement and then things level off. Eventually records are being broken by only minuscule amounts and an approach to the limits of human performance is occurring. At present the record for the 100-meter run is held by Usain Bolt at 9.58 seconds. In 1891 the record was 10.8 seconds by Luther Cary (Guinness World Records, 2019). In a span of almost 125 years the record came down 1.22 seconds. It is almost inconceivable that another 1.22 seconds could be shaved from the current record. A race as short as the 100-meter run perfectly illustrates the concept of limits. At some point it will be physically impossible to go any faster.

Contrast the learning curve or the performance curve provided in Figure 1.1 with that of curve for the cumulative distribution, Figure 2.2 shown on page 34.

It will be noted that the two curves look exactly alike. This might appear to be an astounding coincidence but in fact it is nothing of the sort. In mathematics, the curve or function that occurs above is known as the ogive or ogival function. The standard curve produced under normal circumstances tends to always look the same.

What Moore observed and what others have taken as support for the idea of perpetual growth and progress is an illusion created by viewing only a portion of the curve. Apply this to the actual objects upon which Moore based his initial statements. In the first portion of the curve, the long flat tail to the left that theoretically goes back to the conception of computing, nothing was happening for quite some time. Charles Babbage conceived of the Difference Engine in 1821, and Ada Lovelace wrote what is now recognized as the first computer program in 1842. Despite these influential and

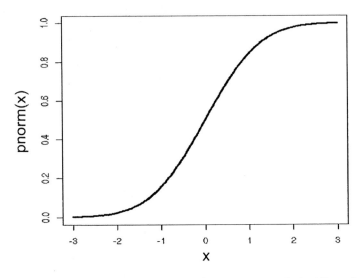

Figure 2.2 Cumulative Frequency Distribution Curve—Ogival Function

seminal first steps nothing resembling an integrated circuit was built for more than 100 years. Enter Gordon Moore who observes the curve at the inflection point as it begins to tilt upwards and accelerate. Only viewing the middle section of the curve gives a false impression that this growth and rapid acceleration is the standard state of affairs, i.e., is a "law." Yet all curves eventually reach asymptote and level out. Moore himself correctly recognized this when he pointed out the fundamental limits of physics upon size. In effect, no transistor can ever get smaller than the subatomic world and as a result, at some point Moore's Law must necessarily break down.

The doubling of processing power is more accurately viewed as an illusion created by a lack of perspective and long-range vision. What seems to be exponential growth is really nothing more than the cumulative effect and the ensuing approach to limits partially observed. This misunderstanding that technological progress is exponential and never-ending has the unfortunate effect of leading to disregard for any destruction caused by innovation. There is no need to worry about the future when new breakthroughs are just around the corner.

Beyond the specifics of Moore's Law to transistor development, there is clearly evidence that demonstrates a slowing of innovation. Absolute numbers of inventions will continue to increase but the pace of innovation, when taken as a marker of progress, can be and is misleading. Although the pace of invention seems to resemble a rapid explosion as evidenced by both the application for and the granting of the issuance of patents, examination of the rate of increase yields a completely different picture.

This is one of those esoteric issues that tend to obscure a rational interpretation of events. Inventions are being produced in greater number, but the rate of this increase is slowing. The upward slope of the curve is approaching the new inflection point. Economist Robert Gordon, in *The Rise and Fall of American Growth* (2016), argues that 1870–1970 were the years of peak growth and that the current era pales in comparison. Using measures of economic gain, he demonstrates that economic growth tied to technological expansion occurred primarily as a "big wave" from about 1850 to 1950. During this period, he contends that a few "fundamental innovations" caused most of the change. Economic growth curves provided by Gordon closely approximate the cumulative curve (he presents the information in a slightly different format using histograms, but the effect and similarities are clear) and demonstrate a dropping of gains as measured by percent increase in GDP gains. A quick summary of invention over the recent past emphatically makes this point.

Table 2.1. Significant Inventions Compared by Time Frame

Pre-History	*Ancient History*	*1000–1850* CE
Fire	Wheel	Movable type
Stone tools	Metallurgy	Water systems—Incas
Lever		Air conditioning— Red Fort, India, c. 1600
		Refrigeration—1830
1850–1950		
Automobile	Radar	Television
Indoor plumbing	Electric lights	Radio
Airplane	Washing machine	Microwave
Telephone	Computer—early	Atomic bomb
Ball point pen	Concrete	Clothes dryer
1950–Present		
Computer—desktop	Smart phone	Internet
Silicon chip	GPS	

In an obvious manner the most significant inventions since 1950 can be viewed simply as extensions of and improvements upon technology developed prior to that time. The smart phone is really just a combination of the phone and a computer while the modern desktop has its origins firmly in the past. Only the Internet stands out as a uniquely novel invention of the current era.

Fundamental change cannot logically be expected to occur at a sustained pace. It might be unimaginative to believe that most of what there is to be invented has already been invented but it is unrealistic to assume that unlimited growth and increasingly significant and impactful innovations will continue indefinitely. What if an invention as impactful as fire or electricity or computers occurred every day? Whether humans could cognitively handle such change is doubtful. Philosophically the issue of whether never-ending sustained growth is a good idea and what the consequences of it might be must be raised.

Growth for the Sake of Growth

Growth for the sake of growth is the ideology of the cancer cell.
—Edward Abbey

For what purpose is technological progress necessary? It might be inevitable that individuals will continue to have ideas concerning various innovations and inventions and seek to bring them to fruition. However, just because it is inevitable does not mean that it is necessary or beneficial. Directionless, thoughtless development, with no purpose other than to produce a commodity that can be sold to make a profit, is not a guarantee or even a likely path to human betterment. What is more likely to occur is a stream of unforeseen and unintended consequences.

Indeed, Adam Smith predicted that changes in economic productivity would eventually lead to a world without work. That seems to be the goal of most technology as well—no *work* but more *things*. Is that progress?

Probably few people reading this ever saw their grandparents exercise. They were too tired from working. The elimination of work may have unintended negative health effects. Children are eating fewer calories, but weight is up, and the reasons are obvious. Leaving aside the issue of whether a world without work is possible still leaves the question of whether it is desirable.

One significant hurdle to unlimited development in the Digital Age is the enormous increase in the need for energy to power the technological enterprise. It might be that the ultimate limit on technological progress, as now defined, is the rapacious need for the energy necessary to power all the devices and machines that permeate the whole of the world. This limit to energy and other resources is hardly acknowledged. Yet everything needs energy. Even things that we formerly manually operated are rapidly being converted. One expression of this might be the electric trashcan.

Even when the subject of the runaway consumption of energy is broached, the discussion usually is given short shrift with a merry "we'll come up with new technology" shrug. This belief in technology to fix all problems, to make "progress," is akin to magic beans for the digital age.

Once you believe you can fix something, you are not afraid to break it. Unfortunately, that is the operating mantra of the digital age—new technology will fix any and all problems. This mentality allows the ability to ignore problems or accept poor solutions in the present and look to the future to get it right. In short, technology will save us.

Technology Will Save Us: The Road to Utopia

More than 40 years ago, Bernard Gendron, in his seminal text *Technology and the Human Condition* (1977), put forth an overt and explicit view of technology as a mythical savior. Written during a time when automation and digital technology were first beginning to emerge he lamented the notion that there was little support for the notion of technology as the guide for humanity. Delineating what he calls the "Best Utopian Argument" for the championing of technological development, Gendron attempted to make a case for technology as the answer to all problems in the human condition. This argument consists of four basis premises that underpin the utopian position of technological supremacy. They are (according to Gendron):

Premise 1: We are presently undergoing a postindustrial revolution in technology.
Premise 2: In the postindustrial age, technological growth will be sustained.
Premise 3: In the postindustrial age, continued technological growth will lead to the elimination of economic scarcity.
Premise 4: The elimination of economic scarcity will lead to the elimination of every major social evil.

In the Utopian Argument, Gendron is mostly talking about mechanized society. Computers were a relative novelty and the Internet was nonexistent at the time he posited this argument, yet the implications for the Digital Age are clear. Indeed, Gendron noted that several thinkers including Norbert Weiner, Daniel Bell, Zbigniew Brzezinski, and Kenneth Boulding were, in the 1970s, already discussing the next revolution in technology that has since come to pass. He noted that "The distinguishing marks of this postindustrial revolution are the emergence of computer technology, rapid development in electronics, communications, and behavior control, and the widespread use of 'intellectual' or 'software' technologies..."

(Gendron, 1977, p. 19). As far as this point goes, Gendron was prophetic. It should be noted that Gendron was not completely convinced of a totally positive nature of technological innovation. He also noted that there was an opposite view of technology, called the "Dystopian View," that bore consideration. This view has become largely the object of ridicule in the current era. Those who advocate caution and regulation are seen as "enemies of progress."

Belief in the utopian view of technology seems to have taken over societal consciousness. In fact, to find balanced views that question whether technological advance is indisputably positive one has to reach back to Gendron and others who were debating these ideas at the beginning of the digital revolution. During the time that has passed since Gendron's initial presentation of the two diametrically opposed views of technological development, technology has insidiously crossed a line from being intrusive and obvious to an almost unnoticed ubiquitous presence due to seamless integration. Voices of dissent have been relegated to the back benches since that time and the world has become infatuated with devices. The Utopian position of the late 1970s has now become the only position acceptable. Despite the near universal acceptance of the Utopian position, the evaluation of this view of technology is not nearly so clear-cut. The evaluation of new technology has become lost in the moment. No longer is the question "should this be done?" but rather "can this be done?" In effect, Gendron's first premise of the utopian position has already been fulfilled.

It is difficult to argue with the first premise of the utopian argument. It appears obvious. Indeed, it is not even necessary to make the argument that a technological revolution has occurred. The evidence is all around. However, examination of the other premises prompts a different outlook. The second premise, that technological advance will be sustained is nothing more than a foreshadowing of Moore's Law, shown to be a misunderstanding and misapplication of a specific observation to a wide range of ideas rather than a factually accurate representation of the future. The collapse of support for this premise clearly undermines even the possibility that the third and fourth premises could occur.

However, even if Premise 2 is accepted for the sake of discussion, Premise 3, that technology will eliminate economic scarcity, is not obvious at all, and a case can be made that it is downright wrong. This premise falls squarely into the realm of blind faith rather than reason. Rather than eliminating scarcity, technology is causing information to be treated as a commodity. Those with access to resources, whether they are intellectual or financial, are at a decided advantage. Technology is not a cure for greed and avarice; it is merely another vehicle to facilitate its expression. Scarcity is not being eliminated; the accounting methods are all that have changed.

People continue to starve, not because of food scarcity but because of a lack of economic resources to purchase the food.

As fanciful as Premise 3 appears, it is grounded in concrete when compared to Premise 4, which implies that technology will eventually cure all social ills. This idea seems rooted in the notion that money is all that matters. Not even a room full of engineers would reject the idea that human emotions might cause some social ills. Most politicians are power mad, not money hungry.

This is not to deny that positive change or progress has been made in the human condition. However, much of the progress as regards the conditions of humanity, from the ending of feudalism and slavery to lessening of the oppression of women, has little or nothing to do with technology. In fact, much of the plotting and scheming and secretive actions that occurred in and around the struggle against slavery likely would not be possible in the modern age. In this respect technology would have likely hindered some of the advances of the past. Tracking chips, had they existed at the time, would almost certainly have been implanted in slaves. Listening devices would have interfered with the planning of almost all social movements from women's suffrage to organized labor. These changes, these substantial advancements in the human condition, are far more attributable to social change than technological advance. Even in the case of the "Arab Spring," the role of technology was likely far overstated. Social media provided some assistance but the conditions on the ground were the culmination of decades of frustration and political maneuvering. The emerging role of social media as a source of disinformation is beginning to overshadow its use as an organizing tool.

Technology has served to uncover and document atrocities and genocidal activities as well as to democratize knowledge and information. It has also served to democratize violence through the widespread availability of guns, explosives, and other weapons. What only nations formerly had the capacity to do; almost any individual, with enough knowledge and will, might now be capable of doing. War is easier; the potential for mass destruction is greater. Technology has given humanity the ability to engage in wholesale slaughter on a grand scale. Indeed, we now have the ability to completely annihilate ourselves with the weapons we have created. Who needs an asteroid, when you've got a brain?

One of the fundamental mistakes often made when conceptualizing technology is to think of it as a solution to a problem rather than as a tool to solve a problem. The analogy of a computer to a pencil is apt and demonstrates the faulty way technology is approached. If technology could solve problems then a pencil would be able to do math problems all by itself. With a pencil this is obvious, with higher tech, there is a tendency to confuse tool and "solution." The more complicated the technology the more likely it is to be elevated to solution.

Technological innovation is often touted as the way to solve a problem for which there is no apparent solution. The burning of fossil fuels is of major consequence to the environment and the technology that has created the problem has evolved very slowly. Even though the problem was created by technological innovation and despite the fact that current improvements on the horizon only minimally impact the problem (and do nothing at a macro level due to the continued increase in both drivers and cars) there is a belief that we'll soon be driving around pollution-free because of innovation.

The problem is more fundamental than can be solved with simple improvements in fuel efficiency and technology. Again, technological innovation is not a solution to the problem of fossil fuel consumption, it is merely a tool to deal with the problem. The real issue is the rapacious consumption of energy and the increasing demands for more. The real answer, *the* solution to the problem of energy demand, is not to find a technology that produces limitless energy. The answer is to stop the demand.

That would require fundamental change in the way people live. That is not going to happen if we pin our hopes on pie-in-the-sky magical solutions to problems. In effect, the belief that we can develop a technological solution keeps us from addressing the problem. An unfortunate type of cognitive bias exists in humans that may prevent seeing the folly of relying on future technology to fix problems. In a study involving problem solving, it was determined that when a solution to a problem is known, it serves to prevent seeing an even better solution (Bilalic & McLeod, 2014). The idea that technology can solve every problem and that future technology will be perfect serves to prevent seeing actual solutions such as conservation.

It is not just that there is a belief in the ability of technology to solve some problems, technology has become the de facto solution for all problems. No matter what, no matter how massive, no matter how intractable the problem a perfect technological solution is always deemed to be just around the corner. The problems that technology might produce are inevitably ignored or minimized. Even perfectly good devices, those that work as intended, are constantly being improved, upgraded and rendered obsolete. Every minor advance is heralded as a breakthrough. A degree of the planned obsolescence feature of many modern devices is without a doubt a product of marketing and the profit motive. However, it is not at the level of minor gadgets and their price that should cause alarm. Instead the real issue lies in the unquestioning acceptance of technology as progress no matter what the actual contingencies. The inherent danger in many technologies appears to either be ignored or subjected to cognitive distortion.

This distortion serves to minimize any inherent danger by relying on technological safeguards. We tell ourselves not to worry about the machines because they are safe. It's like sleeping in the room with a crocodile but not worrying because it's chained up. It comes down to how much you trust the chain.

The history of the development of weapons is illustrative of the problems of using one technology to fix the problems created by an earlier technology. It can be summed up thusly: Every new weapon was supposed to be so horrible that it would make all other weapons and by extension, war, obsolete. Instead, we have moved from the rock to the atomic bomb. Instead of being safe, we now worry about crackpot ideologues, rogue terrorists' cells, and even unstable governments blowing up the world. Failure to address the underlying aggression in the human species will not be solved by technology, it just ups the ante.

Global warming is another area where belief in the ability to fix the problems prevents the taking of appropriate action. Climate change deniers often minimize the effects of climate change by pointing to a future when we will figure out how to reverse the effects without changing the way we live. In effect, technology will save us. That does seem to be the mindset. All new technologies carry the promise that they can fix the problems of the last technological innovation. In fact, there seems to be no limits to the trust in technology to rescue the future. Despite our own experiences, the belief that technology can fix everything is ubiquitous. Technology, we are told, can even prevent us from dying and allow us to live forever.

Table 2.2. Death by Progress: Problems Created by Technological Advance

Technological Advance	Problems Created
Radium	Deaths of scientists and workers in watch factories
Nuclear power	Radiation/mutations/annihilation
Computers	Resource depletion/Child labor
Weapons of mass destruction	Wholesale death
Asbestos	Asbestosis and mesothelioma
Thalidomide*	Birth defects
Lead	Lower IQ, cognitive deficits
Internal combustion engine	Climate change

*Thalidomide is just one of a multitude of teratogens and iatrogens produced by technological means that have caused birth defects. Lead and mercury are perhaps the best known but by no means only materials implicated in various human health problems.

Resource Depletion: Alienation from Nature and Magical Thinking

The emergence of new technologies will wean us of our dependence on foreign oil.
                                                          —George W. Bush

There is likely no area where the tendency to look for technological solutions is greater than in the area of resource and energy consumption. Alarm about the growing consumption of energy and other resources tends to focus on solutions geared towards producing more power or smart devices that make small incremental improvements in usage rates. The simplest solution, turning off devices or not having them to begin with, is rarely broached.

The Union of Concerned Scientists has put forth a 20-year-plan, *Half the Oil*, to significantly reduce energy consumption (Heid, 2013). At the heart of this plan are measures aimed at conservation. Because it relies on no new technological breakthroughs the plan has gotten little attention. Instead, promises of "clean" coal and plans for carbon sequestration are offered that fail to address the central issue of consumption. Never mind that an international coalition of scientists who are experts in the area are calling for drastic changes to the way humans live. The belief that there will be new inventions that allow humans to continue burning fossil fuels at ever increasing levels without consequences seems widely accepted.

Were this belief not widely accepted it is difficult to believe that much of society would operate in the manner it currently does. All industrial and digital society is geared towards the production of devices that require energy. Every device requires charging and the push is on to produce more and to place the goods of technology in the hands of everyone. There is little concern that the energy might not be available.

Water resources are equally squandered. Shortages in areas other than natural deserts are almost solely related to human activity and the use of technology to redirect the water. Corporate farming is a major culprit as it is estimated that 70 percent of the world's fresh water is used in irrigation by the agricultural business (Global Agriculture, 2015). The Colorado River no longer reaches the ocean, primarily as a result of irrigation and municipal use (Zielinski, 2010; Howard, n.d.).

Desalinization, the process by which fresh water is produced by the removal of salt, minerals, and impurities, is heralded as a means to allow humans to live where they should not. Many nations in the Middle East depend upon this technology for fresh water. Naturally, drought-stricken areas around the world are beginning to seek this technology as the answer to their problems. Discussion of this topic follows that of many others in that a technological solution is proposed to a problem largely or wholly created by technology. Undeveloped technology will fix any unforeseen problems is the magic thinking.

Environmental concerns are almost always brushed aside as "too costly" before any actual analysis of costs can be made. All solutions are on the table except conservation. Meanwhile, every other house in Arizona seems to have a swimming pool, every child's toy needs batteries, and every office building in every metropolis keeps the lights on all night. Hyperbole to be sure, but not far from the truth. Fast food convenience outweighs the collective concern for the welfare of the planet. Saving energy seems like too much work. Keep that thermostat cranked rather than put on a sweater. Better to have scientists split atoms than make any effort to change.

Nuclear power is the ultimate technological delusion. A basket of fish to feed the multitude. A reactor full of plutonium and limitless energy for free. It perpetuates a belief in safe, clean energy at little to no costs. This idea bears no resemblance whatsoever to the realities of nuclear power, which is expensive and filthy, and has a high potential for disaster. Even if energy could be produced for free and even if there could be assurances of perfect operation, there remains the problem of waste in the nuclear landscape. Fuel rods necessary for the operation of nuclear plants may pose their greatest hazard. With a half-life of forever, disposal is not just a matter of dumping them; it is a matter of preventing them from being disturbed—*forever*!

Even if safety issues could be addressed at a level that is fail-safe, the fuel rods themselves will continue to be a problem. Even if it were possible to find a place to store them that is completely isolated and seismically stable, it is not possible to ensure that the conditions under which the rods are buried will remain the same. It is not just a matter of ensuring that the conditions under which they are initially buried are safe, it is also a matter of everything remaining the same. Not only must seismic conditions remain the same (plate tectonics?) but so must population patterns, geopolitical situations and even the will to prevent tampering. It is highly unlikely that the world will have the same political alignments for the next 5000 years. Is it even remotely possible that there will be a smooth transition of governments over millennia that will result in preserving the status quo? It is more conceivable that society will lose track of where the rods are buried than that there will be no geopolitical change in that span—and, if the half-lives of the material in question is considered, forever? The costs to those who stumble upon these buried rods of death will be far worse than the curse upon those disturbing a pharaoh's tomb; it could prove to be a catastrophe for an entire society. Burying radioactive fuel rods is nothing more than a time capsule of death.

The belief in technology to reverse the damage it has created seems to be nothing more than a massive case of delusional thinking. It is as if society is collectively saying, "It will be okay, it will be okay. Technology will save us." Belief in technology has become the new religion. Rational thinking is forbidden. Better living through technology has become the dogma of the belief system. A god has been made of technology and has taken the place of critical thinking.

Outsmarted by a Virus

In stark refutation of the supremacy of technology to save us and solve all of our problems stands the reality of the corona virus pandemic. Often referred to as the "novel" (i.e., newest) corona virus, this pathogen is one of several viruses in a family of related viruses. Likely emerging from bats (Del Rio & Malani, 2020), this new pathogen is related to MERS (Middle East Respiratory Syndrome), SARS (Severe Acute Respiratory Syndrome), and a few variants of the common cold; all are classified as coronal viruses. Novel corona, also designated Covid-19, or more accurately SRS-CoV2, belongs to the same family.

This virus is almost perfectly suited to spread and kill. There is a relatively long incubation period (up to 21 days) and many carriers, by some estimates up to 50 percent are asymptomatic. Many others with the virus experience only mild symptoms and may not accurately interpret these as a sign of illness. Because of these factors, the virus has the ability to completely penetrate a population before anyone knows it is present. Recent theories have also suggested that those most likely to be contagious are those most likely to continue with normal behavior and eschew social distancing (Barton, Bennett, Cook, Gallup, & Platek, 2020). The outbreaks in Louisiana following Mardi Gras, those in Florida associated with Spring Break, and even the initial outbreak in China associated with New Year's celebrations are but a few examples of this phenomenon.

Younger people apparently have milder symptoms in general, making them ideal vectors. With the young being either asymptomatic or experiencing mild symptoms they then can spread it more easily through the older population. Children to grandparents, who are in the age group most affected, becomes a lethal vector. Some of the best evidence as to the insidious nature of the virus was the realization that the virus had likely been in southern California as early as November, 2019 (Anderson & Smith 2020), while other evidence suggests New York may have had over 10,000 cases before the first case was identified (Carey & Glanz, 2020).

As with many other problems, technology has promised solutions to

Covid-19. There is clearly a belief that a vaccine of treatment will emerge. This belief remains even though the common cold, one variant of coronal viruses, is still without a vaccine or treatment despite decades of development. SARS, with a mortality rate of approximately 9.9 percent was eventually eliminated through public health measures. MERS continues to hang around, with 2519 cases and 866 deaths (fatality rate approximately 34 percent) reported in 2020 according to WHO (NAIAD, 2020).

With Covid-19, the impression is that we are pulling out all the stops to find a solution. The ultimate in high tech, artificial intelligence, is being employed to run genetic sequencing programs to find a means to disrupt the virus (Wu, 2020). In addition, numerous field trials of medications are under way, and tracking apps are being developed to assist contact tracing (Griffin, 2020), while drones are being employed to monitor temperatures remotely (Dormehl, 2020). It is entirely possible that some of these methods being deployed will assist in controlling or curing the disease. The possibility that a cure will not be found must also be considered. The development of vaccines and treatments for disease is not an assured process.

We already know what works—tried and validated public health measures. As noted, SARS was eliminated by way of these same public health measures. Ebola outbreaks in Africa have not been controlled because of medical breakthroughs but rather because of public health measures such as containment and contact tracing.

Contact tracing is a labor-intensive process of tracking every infected person's contacts and then all their contacts and so on until you identify all the infected. This is an old-style public health approach that has been employed for more than 100 years. Technology through means such as tracking apps could be of assistance here, but the basic process involved has nothing to do with technology. This is yet another place where a confabulation of science and technology leads to bad decisions. Science, through data, will lead to the ability to determine what the effects of the virus are, how widespread it is, and how to approach combating it. Technology offers a means to assist this process at times, but also promises a solution, whether it is deliverable or not. Vaccines developed for Covid-19 may eradicate it worldwide, but the probability of this is anything but certain. It may not be possible. It may be that like the common cold, a cure or vaccine remains elusive.

This reliance and emphasis on technology to find a vaccine or cure is to some degree misplaced. Meanwhile, what has worked has been anything but high tech. What has been effective at preventing the spread has been social distancing and staying put. For those who must come into contact with others, again the most effective methods have little technological imprint. Barriers, such as masks and other personal protective equipment (the fabled PPE), are the best methods to prevent infection.

Science, to a degree, predicted this outbreak. Concern for an outbreak of a disease related to corona virus in bats has been noted for nearly 15 years (Lau et al., 2005). Two peer-reviewed publications in 2019 practically screamed about this possibility. One article, with the title "Viruses in bats and potential spillover to animals and humans" (Wang & Anderson, 2019), provided a trenchant warning of what was to come while the other went so far as to identify what species of bat was most likely to produce a coronal virus infection that might cross into humans (Lau et al., 2019). These studies also bring us back to the point above, that efforts to address the effects of coronal viruses have proven difficult and the notion that a cure is around the corner may be nothing but magical thinking.

An interesting dichotomy has emerged as regards science and technology associated with efforts to control Covid-19. The science side of the equation appears to have some handle on the virus. Science has allowed for the calculation of illness trajectories and has through public health measures effectively made the case that social distancing is effective in controlling the spread of transmissible diseases.

Technology, by contrast, has made a lot of promises that have not been delivered. At the beginning, the CDC first tests were faulty and caused setbacks. New tests have been plagued by high, or even excessive rates of both false positives and false negatives. The supply of tests, while certainly impeded by governmental nonsense, has largely shown the failure of a health care system based on profit.

On-demand, no-inventory supply chains, the kind hailed as a technological marvel, have come to pose one of the biggest threats associated with the virus. Outbreaks in the virus among the workers in multiple food processing plants have led to significant disruptions in the food supply. By aggregating animals and turning farming into a manufacturing process, a vector of infection for the employees and then the community is created. The alleged flawless, waste reducing, on-demand supply chain managed by Big Data has led to catastrophic shortages of critical supplies in various sectors from food to medical equipment.

Of course, it is in the medical arena, where the critical shortage of personal protective equipment and ventilators has been among the greatest concerns during the crisis. In this area, the most effective equipment that of a full "balloon" suit, in which the health care provider is completely shielded and encased in a pressurized suit, is nearly nonexistent primarily because of costs of production and maintenance. It must be noted the disgrace and utter failure of a health care system that bills $12 for a cotton ball or band-aid but cannot successfully provide equipment to protect its workers. Instead, what results is an overwhelmed medical system where workers are left to their own devices to acquire PPE, and doctors are improvising ventilators out of parts from hardware stores. Concerned people are producing masks in their homes for workers, and governors are competing with each other on the open market to acquire medical equipment.

A demonstration of the failure of technology to solve this pandemic can be seen in tracing the spread of the virus. The incidence of the virus is greatest in the most technologically advanced countries while it appears to lag in attaining a foothold in nations that are less technologically advanced. The primary feature associated with the spread of Covid-19 is connectivity. Flash animation at the Johns Hopkins University Coronavirus site (2020) demonstrates this with clarity. The virus spread through nations with connections to each other. Nations with high rates of exchange had the highest and fastest rates of penetration. Airplanes, without a doubt, were a primary culprit in the rapid spread of the disease. One of the reasons the U.S. fared so poorly at the beginning of the pandemic was the web of connections. Not only was the country connected to other countries, the U.S. has multiple hubs of connectivity while most countries have only one or two.

However, in the case of the U.S., it must also be noted that the incompetent governmental response hindered any and all possible responses. From the failure to adequately provide testing and testing materials to the insistence on a free market approach to the medical supply chain, the U.S. has provided a template of how not to do it The incompetence of the federal response to the pandemic has been outlandish but does not compare to the unfortunate lack of leadership on the matter. It would be nearly impossible to catalog the times Donald Trump misstated information, distorted the facts and generally dissolved public confidence in the federal government's response to the crisis. From calling the pandemic a "hoax" to suggesting that people ingest disinfectant, nothing of use has originated from the top. Unfortunately, it is the complete undermining of science and the medical community in service to propaganda and conspiracy theories that has proven to be one of the greatest impediments to coordinating an adequate response.

The political divisions in the U.S. have clearly exacerbated the effects of the virus. The ability of people to trust information now goes through a filter that includes the source. Most of those making decisions—i.e. politicians—have little background in or understanding of science. As a result, belief in magical cures and erroneous information proliferates.

The internet assists in the undermining of a coordinated response around the nation and the world as conspiracy theorists and paranoia abound. There is no end to the theories circulating online, including everything from 5G to Bill Gates as culprits. Those who feel that being socially responsible and staying at home is somehow part of a government conspiracy to take their guns have taken to the streets in several cities, with their guns, to demand the country be reopened.

Unfortunately, such lunacy has repercussions. The governor of Georgia, Brian Kemp decided to "open" the state on April 24, by allowing barbershops, nail salons, massage parlors, tattoo shops, and gyms to go back to business even though public health officials warned against this action (Judd & Bluestein, 2020). With this move Georgia embarked on a trail to demonstrate what happens when stupidity reigns.

This action in Georgia will create an opportunity to examine the effects of social distancing by the time this book is published. On April 22, 2020—Earth Day 50—both India and the state of Georgia had roughly the same number of cases at around 20,000. India extended its lockdown while Georgia opened the floodgates. Following this out for one month, six months, even a year will provide a measure of the effectiveness of staying put compared to re-opening the economy. It should be noted that India is at a distinct disadvantage in this comparison with a population of 1.3 billion, roughly 130 times that of the Georgia. India has civil unrest, mass numbers of migrant workers with no place to go, and a public health system that is spread thin. Compounding these problems is the fact that social distancing is nearly impossible in India. While India presents the opportunity for an extreme comparison, the experience of Georgia as regards all other areas will be important to reckon.

One theory that appears to accompany the outbreak of all infectious pathogens is the idea that it was created by humans. In effect, the idea that one nation or another has engineered a virus for nefarious purposes has been associated with Ebola, Marburg, MERS, SARS, and other contagions over the years. Although Covid-19 was predicted by scientists, and the virus at present does appear to have originated in bats (Del Rio & Malani, 2020) there continues to circulate propaganda that the virus was created in a lab in China. The fact that the Wuhan CDC is less than 400 yards from the "wet market" (live animals for sale as food) associated with the first case, lends further fuel to such conspiracy theories. At some level, even the existence of the wet markets around the world demonstrate the failure of the ability of agribusiness to adequately feed the world. Crop yields may have been increased but the failure to feed people contributes to the need for places such as wet markets.

China has not helped, claiming at one point that the source of the infection was a U.S. sailor, and removing research reports about the virus. Eventually we will likely figure out the original source of the infection and all evidence presently known will need to be reviewed. It may be that the source is not at all what initially appeared to be the case. The origins of the "Spanish" flu pandemic of 1918–1920, is still debated among researchers with some believing it may have originated in Kansas (Barro, Ursúa & Weng, 2020).

The murkiness surrounding the identification of the origins of such diseases also serves to fuel paranoia and propaganda efforts. Add to this the knowledge that various nations actively maintain labs dedicated to research on deadly pathogens and it is easy to understand why there is so much distrust of official accounts. Blatant efforts by governments and their officials to distort reality and manipulate public opinion exacerbate efforts to determine basic scientific information. When the U.S. president suggests at a briefing for people to drink disinfectant or to take off label medication, it is difficult to trust even basic health directives (a day or two later he claimed he was being sarcastic).

Science tells the truth; technology offers false hope. Science tells how bad and offers a warning. Technology assures that all will be okay and gives false assurance.

In a bizarre twist to the pandemic, there appear to have been positive environmental effects of the worldwide lockdown. Oil futures fell below \$0 for the first time ever and demand became nonexistent as nearly everyone quit driving (DiSavino, 2020). CO_2 levels almost immediately fell by detectable amounts and the air around the world visibly cleared as a result of reduced human activity (PBS, 2020). These gains will almost assuredly disappear once activity resumes, but the point is made as to the effects of human behavior. In many ways this could be a template for reducing emissions, but this opportunity will likely be squandered, as a growing economy is seen as even more sacrosanct than technology.

Interestingly, in a study from northern Italy, pollution was found to be a co-factor in risk for Covid-19 (Conticini, Frediani & Caro, 2020). It does not seem to be a startling revelation that a respiratory illness would be exacerbated by dirty air, but it does bear repeating that human activity serves to exacerbate many health problems. The benefits of a clean environment are not just aesthetic but critical for the health of all the organisms on the planet.

The emergence of corona virus from bats living hidden in caves would be surprising were it not for the fact that humans, with the assistance of all manner of technology from large machines to nanobots, are invading and destroying the natural world. Several of the more lethal pathogens that plague humanity, Ebola, hantavirus, Marburg, etc., all emerged in places that have been remote and almost beyond the reach of most human beings. It is excursions into these areas that have unleashed these catastrophes. As humans invade the far reaches of wildlife, encounters with unknown pathogens should be expected to continue. Imagine a virus with the lethality of Ebola but as contagious as Covid-19. Technology would have no response. Medical facilities would be decimated and likely avoided at all costs.

3

Killing All the Bees

Technology and Environment

If the bee disappeared off the face of the earth, then man would have only four years of life left. No more bees, no more pollination, no more plants, no more animals, no more man.
—Maurice Maeterlinck, *The Life of the Bee*

The technological revolution of the past 400 years has had what can only be described as a profound effect on the natural world. This effect can be seen as one of degradation and destruction of the environment from the largest macro effects upon climate to the destruction of microorganisms. Illustrative of this impact are the current circumstances of the honeybee.

Bees serve as the primary pollinators of the vast majority of foods eaten by humans. According to the National Resources Defense Council (NRDC, 2015) bees pollinate 71 of the 100 major food crops, accounting for 90 percent of what is consumed. The USDA (2015) reports that bees pollinate one-third of all fruit and vegetable crops and that many crops (e.g., apples, peaches, blueberries, onions, broccoli, carrots, etc.) are almost completely pollinated by honeybees. The United Nations (2011) estimates that bees add over $200 billion to the world's economy through pollination of crops. To demonstrate the effects of honeybees upon food crops, Whole Foods grocery removed all the foods on its shelves pollinated by bees. This action resulted in 237 of 453 items being removed from the store shelves (Whole Foods Media, 2013). In short, for those who like to eat, the bees be it.

Unfortunately, bees are in trouble. A condition known as colony collapse disorder has been devastating the bee populations of the world. Research conducted jointly by the USDA, the Apiary Inspectors of America (AIA) and the Bee Informed Partnership (http://beeinformed.org) highlights the severity of the problem. Since 1947 the number of honeybee colonies in the United States has declined from approximately 6 million hives to around 2.74 million. This represents a decline of more than 50 percent in that time frame. Unfortunately, the vast majorities of these losses have come in the recent past. In 2012–13 slightly over 45 percent of colonies lost. Several states (Illinois, Oklahoma, Iowa, Michigan, Pennsylvania, Maryland, Delaware, and Maine) reported losses of more than 60 percent. In the 2014–15 survey period 42.1 percent of managed colonies were lost. For 2015–16 the overall loss rate was 44 percent of colonies with 28 percent lost during the winter when rates are generally lower (Seitz et al., 2016; Steinhauer et al., 2016; Kulhanek et al., 2017). In the 2017–18 year there was noted to be some improvement with 33 percent lost, although some states had rates that were catastrophic. Arizona lost 72 percent, Tennessee had a 65 percent rate and Louisiana lost 60 percent. Eight other states had rates over 50 percent (Bee Informed, 2019).

When colony collapse disorder first emerged, it was a mystery to beekeepers. Healthy hives would be producing honey and apparently thriving one day and completely gone the next. Beekeepers knew something was going on but were unsure of the cause. Beekeepers have long dealt with a host of problems affecting bees, from the *Varroa destructor*, a mite that lives in the throats of bees, to the *Nosema ceranae*, a fungus from Asia that plagues colonies in the United States. Colony collapse was something entirely different. Entire colonies were wiped out overnight.

At first researchers did not know where to look. Several theories emerged, including the idea that cell phone towers or other sources of electromagnetic radiation might be disrupting their ability to navigate. Pesticide use and habitat loss were also possibilities that were investigated. Genetically modified crops also played into this equation in that what most are modified for is to withstand pesticides. Monoculture, as practiced by corporate farming, also was investigated as a culprit, in that it removes diversity from the biosphere.

Science has now closed in on the cause. According to a study conducted by the Harvard School of Public Health (Bennett, Bellinger & Birnbaum, 2016) a class of pesticides known as neonicotinoids or "neonics," are the likely culprits. Bees come into contact with these pesticides as a consequence of the dust contaminated with neonicotinoids that is released into the air during planting and application (Krupke, Hunt, Eitzer, Andino & Given, 2012). These chemicals have been found to affect the ability of bees to navigate and to suppress the immune system as well. The immune suppression effects are particularly problematic in that they also tend to produce conditions that make bees even more susceptible to the mites and parasitic funguses that already plague bees. The European Union banned the use of these chemicals for a two-year period (Dunmore, 2013). The EPA declared neonicotinoids harmful to bees in 2016, and the U.S. Fish and Wildlife Service announced it will phase out the use of all neonicotinoids on all of the

150 million acres of the National Wildlife Refuge System (Friends of the Earth, Annual Report 2014).

Almost as soon as one problem disappears another seems to occur. Research has now begun to point to glyphosate, an herbicide that has also been implicated in several health conditions including Parkinson's and autism, as another factor in the destruction of bee colonies. Glyphosates disrupted the gut microbiome of healthy bees, making them more susceptible to the panoply of other problems that affects them (Motta, Raymann & Moran, 2018).

Other pesticides are also likely having effects. The USDA APHIS Survey (2016) found 120 different pesticides in 1,078 samples. The samples were taken from pollen stored inside the brood comb, the innermost section of the hive, and reveal the extreme exposure to which bees are subjected. Approximately 50 percent of all hives tested positive for three or more pesticides and one hive was found to contain traces of 21 different pesticides (USDA, 2016). Making matters worse, the EPA has begun removing restrictions on certain pesticides such as sulfoxaflor, as a policy and political statement (Knickmeyer, 2019).

While honeybees get most of the attention, wild bee populations that do not produce honey are also important as pollinators. Although far less studied, they also appear to be in trouble. In a survey of 119 wild bee species it was found that 14 species were in significant decline. The survey put the decline rate at 90 percent over the past 125 years (Mathiasson & Rehan, 2019). This same study found shift in ranges related to habitat loss and climate change. Pesticides were also implicated.

The bees may not be the only organisms affected by pesticides. There is also evidence that pesticides may be affecting the olive trees in Italy in a similar manner, serving to weaken the organism in such a way that other threats have a greater effect (Brown, 2015). The European Food Safety Authority has also reported that neonicotinoids "may affect the developing human nervous system" (Bee Action Network, 2015). Clearly, any application of the precautionary principle would suggest that this class of pesticides is too suspect to continue to employ. To prevent and reverse the damage already done to the bees, large-scale changes will almost certainly have to be made. Given the relative political power of corporate agricultural interests, the likelihood that these changes will be made without a fight are nil.

The fact that the bees are being affected is alarming on another level as well. In many ways, this world belongs to the insects. In terms of actual number of species and number of organisms they rule the planet. This is in large part due to their resilience and resistance to most diseases that plague other animals. Everyone has heard the line that the cockroach is the only thing that can survive a nuclear holocaust. Maybe this is true, maybe not.

The only way to really tell will, one hopes, not occur. The point can be easily made, however, that insects are tough. Something is affecting them.

Insects, which also serve as a major food source for many other species, are in trouble. According to a review of studies of insect populations, 41 percent of surveyed species were experiencing serious declines (Sanchez-Bayoa & Wyckhuy, 2019). Many species studied were on the verge of collapse in another survey (IPBES, 2019). The extinction of insect populations will have the severe effect of wiping out their predators as well. Indicative of the level of destruction, the total insect biomass has declined by 76 percent in the past 30 years (Hallman et al., 2017).

The effects of climate change are not limited to declines in populations. Some species, notably the mosquito, are expanding their range and their breeding periods appear to be increasing (Kearney et al., 2009). An interesting side effect of this is that birds appear to be at greater risk of malaria as a result (Garamszegi, 2011).

Beyond the insects and the animals that feed upon them, there are other signs emerging of alteration of behavior among species.

A species of ant from Brazil, *nylanderia fulva*, also known as the "tawny crazy ant," appears to have a particular taste for all things electrical. The ants have been noted to live inside electrical devices and are often discovered when they cause malfunctions. Present in large numbers in Texas and a few other southern states the ants are thought to cause damage while looking for nesting sites (Maron, 2014). However, according to David Oi, a researcher and etymologist with the Department of Agriculture, it cannot be ruled out that the ants are attracted to electrical currents (Mooallem, 2013). Ants feeding on electricity? More likely is that as the ants enter electrical circuits they are electrocuted and send out signals indicating they are under attack, leading to even more ants (Main, 2013). Still, even the remote possibility of ants attracted to electricity is fascinating and alarming at the same time.

Beyond the bees and ants and other insects, animal populations in general are declining at precipitous rates. According to a longitudinal 40-year study carried out from 1970 to 2010, a 52 percent decline in animal populations was observed (World Wildlife Fund: Living Planet Report, 2014). In less than five years this decline had reached 60 percent (WWF, 2018). Beyond this decline in living species, the rates of extinction are also on the rise.

The rate at which species are going extinct has led to concerns that earth is entering a period of mass extinction as marked by the loss of 50 percent of species. In the history of the planet there have been five previous "mass extinctions" as designated by this level of extinction. The possibility that the planet has entered a sixth period of extinction has been submitted by several researchers and authors (e.g., Kolbert, 2014; Billings, 2014,

Leakey & Lewin, 1996). This die off has occurred within the past 50 years. Its significance will be explored in greater depth in the later chapters.

If it is possible that these statistics could be presented in a more grim manner it would be found in a recent laudable effort to measure the total biomass of the planet. This effort estimated that human activity has resulted in the destruction of 83 percent of the mammals and 50 percent of the plant life on earth (Bar-On, Phillips & Milo, 2018). It also reported that fully 60 percent of the mammals on earth are livestock while only 4 percent are wild animals (humans comprise 36 percent). This imbalance of livestock to wildlife is stunning. It is difficult to accept the numbers.

Wildlife is threatened from both the direct and indirect effects of technological development. Birds are also disappearing at an alarming rate. According to a recent investigation, a drop of nearly 3 billion birds, a 29 percent decline, has occurred in the U.S. and Canada in the past 50 years (Rosenberg et al., 2019). Some species such as bees, reptiles and birds are directly threatened by chemical exposures, but the most significant threat to animals is habitat loss. In a report issued by the Rockefeller Institute, at least one-quarter of natural habitats have been converted to other uses. The primary culprit is agribusiness (Whitmee et al., 2015). Other factors such as the clear-cutting of timber, mining of coal and metals, and oil and gas exploration contribute heavily to the destruction of the natural world.

What is being dumped into the water may turn out to be a more significant problem than that of disappearing aquifers. Water is necessary for life. What gets in the water also gets into those who consume it. Animals that live in water are particularly susceptible to these effects. Frogs and other animals with semi-permeable membranes for skin have been some of the first to show problems.

Severe declines in the numbers of water dependent species such as salamanders, frogs, and other amphibians have been noted for some time (Blaustein & Wake, 1995). In addition to declines in numbers, mutations and other problems plague reproduction. Deformed frogs have been found throughout the world with more than 60 different species affected. Alteration of habitat by human activity appears to be the major culprit in these findings (Blaustein & Johnson, 2003). Fertilizer run-off, radiation, and pollution all contribute to environmental factors that then make the animals more susceptible to parasitic infections, which further alter the species. Historical reviews have determined that the rates of abnormalities are increasing at a marked rate (Blaustein & Johnson, 2003). More recent studies have noted the effects of environmental stress as a result of climate change and predict that these problems will continue to intensify (Rollins-Smith, 2017). Habitat loss, primarily through the loss of wetlands, no matter what the circumstances, poses the greatest threat to amphibians.

Animals higher on the food chain are also being affected. Examination of orcas and other cetaceans in European waters have found high concentration of PCBs lodged in their body fat. These chemical contaminants are ultimately threatening the reproductive abilities of the animals to the point that they are threatened with extinction in Europe. Waters around the U.S. have lower concentrations due to earlier banning of PCBs (Doyle, 2016; Jepson et al., 2016).

Multiple chemical compounds are also found in groundwater. From graphene (Bourzac, 2014) to pharmaceuticals (Heberer, 2002; Webb, Ternes, Gibert & Olejniczik, 2003; Jones, Lester & Voulvoulis, 2005) to any number of industrial chemicals, such as flame retardants (Cribb, 2014) and pesticides (PPDB, 2019), the water supply is tainted.

Evidence that people no longer trust their drinking water to be safe can be found in the amounts of bottled water consumed. Yet bottled water is even less regulated than tap water. Numerous comparisons have found that bottled water is often more contaminated than water out of the tap. The fact that most accept or believe that their water is undrinkable should be alarming on a much larger scale than it seems to be.

All these plastic bottles are becoming a separate problem. The giant floating island of plastic goo in the Pacific is one well known result of this problem. Known as the Great Pacific Garbage Patch, it is nearly twice the size of Texas, gathering where the oceans vortices collide. This example is but one symptom of a much larger concern. Plastic, in all its various permutations, is turning out to be an eco-nightmare.

Plastics and Other Toxins

Plastic is a general term for a wide variety of materials that can be easily molded at relatively low costs. The production process essentially allows for this broad class of materials to be adapted depending on purpose. Its primary and overwhelming advantage, however, is not its malleability, but rather its cost. Modern manufacturing would have to completely revamp without this material. It seems ironic that plastic's advantage in manufacturing is thought to be cost since it is essentially composed of byproducts of the petrochemical establishment, which is not generally viewed as a low-cost industry.

Plastic, as it turns out, is everywhere. Like plutonium it also seems to last forever. Since 1950 there has been approximately 10 billion tons of plastic produced with nearly 7 billion tons ending as waste (Parker, 2018). Plastic is mostly a one-use product that immediately makes its way into the environment. It does not disappear as it breaks down but rather degrades into microplastic which becomes even more difficult to remove from the environment. In the Great Pacific Garbage Patch the plastic is barely visible from the surface. Below the surface it is an oozing mass slowly sinking to the depths of the ocean. Microplastics have also been found in ice floes in remote areas of the Arctic and researchers suggest that the area could be a "sink" for plastics (Bergmann et al., 2019).

Plastic has been found in wildlife throughout the planet, from birds to marine life. The ubiquitous nature of this problem for wildlife can be illustrated by two articles so close together in time in the same newspaper with headlines so similar it appears to be the same story. One, from Italy, reported a whale found with 48 lbs. of plastic in its stomach (AJC, 2019, April 2) while the other noted a whale in the Philippines with 88 lbs. of plastic in its stomach (AJC, 2019, March 19). Possibly more alarming, in a study examining all 7 major species of sea turtles, microplastics were found in the bodies of EVERY turtle examined (Nelms et al., 2015). A more comprehensive study found that more than 700 species have been documented to be contaminated with plastics, including 90 percent of surveyed bird species and half of all cetacean species (Fela, 2018).

Humans have not escaped contamination by microplastics. A meta-analysis of 26 studies found human consumption averages of 39,000–52,000 particles of microplastics and 74,000–121,000 particles when the amount inhaled through respiration was included. Those who drink only bottled water were estimated to have taken in another 90,000 particles per year (Cox et al., 2019).

Particularly alarming in terms of the effects produced are substances that fall into the category of endocrine disruptors. Multiple chemicals, such as the now banned Bisphenol-A, related to the manufacture of various forms of plastic are having clear and direct effects upon human health. These chemicals have been implicated in emergent problems such as early onset of puberty, the decline in male fertility rates, asthma, genital deformities, hormone disruptions in children, impaired immune systems, and several other conditions. Many of these problems are related to the leaching of estrogen from various forms of plastic often used to hold food or water. Research into the issue of estrogen leaching suggests that no plastic products are safe. An investigation into nearly 500 different products made from various types of plastics and all reportedly BPA-free, revealed that 72 percent leached synthetic estrogen (Blake & Mar, 2014; Yang, Yaniger, Jordan, Klein & Bittner, 2011).

Unfortunately, plastics are not the only materials that cause severe health effects. The planet is contaminated in almost all corners and the health of every single being is affected by the by-products of the industrial and digital world.

Testing of new compounds is generally believed to be the norm. That is untrue. In fact, as a result of the way in which chemicals are regulated, there is a disincentive to test products for safety. The very nature of the industry and perversely written regulations serve to obscure the severity of the effects of chemicals upon living organisms. In manufacturing, thousands of chemicals are routinely used but are unknown to all but those who use them. When industries develop a process that involves a new non–naturally occurring chemical compound, they do not announce it to the world. More importantly, they do not announce it to their competitors. Rather, the new compound is held as a proprietary secret to be used only by that company. Testing for safety is not even an afterthought. It does not occur.

Before a chemical compound can be tested there has to be some concern as to its safety. The Toxic Substances Control Act (TSCA) requires that a chemical "…present an unreasonable risk to human health or the environment…" before there can be any efforts to regulate it. In other words, chemicals must have a noticeable effect before they are even assessed. At the point where damage becomes obvious it is too late.

Lead and mercury were used for thousands of years before the effects began to be understood. Tobacco took decades to regulate even after its health effects were known. Chemicals such as asbestos and Teflon became regulated only in 2016.

Table 3.1. Environmental Toxins

Chemical(s)	*Damage Caused*
Phthalates	Endocrine disruption
Polychlorinated biphenyls (PCBs)	Cancer
Plastics	Endocrine disruption
Hexachlorines	Cancer
BPA	Early onset of puberty/Endocrine disruption
VOX/SOX	Liver and respiratory issues
Flame retardants	Liver damage
Dioxins	Cancer
Hexavalent chromium, chromium-6	Cancer
Lead	Cognitive deficits
Mercury	Cognitive deficits
Asbestos	Asbestosis, cancer, lung disease
Halogenated chlorofluroalkanes	Cancer
Pesticides	Cancer, neurological damage, genetic damage

SOURCES: Nriagu, 1988; Crinnion, 2000; Ahlborg, Brouwer, Fingerhut et al., 1992; Schwartz, 2004; Lanphear et al., 2005).

Pesticides, as a class of chemicals, are ubiquitous in the environment. The enormity of the exposure worldwide is difficult to comprehend and beyond the scope of this text to fully explore. A Google Scholar search for pesticides will find over 400,000 studies associating the chemicals with negative health effects. However, unlike many chemicals in use, there is a concerted effort to track pesticides through the Pesticide Properties Database (PPDB). This free-to-access database catalogs information on approximately 2500 different pesticides and 700 metabolites. Most of the other substances listed in the table above are difficult to track and their distribution throughout the environment is not remotely understood.

What lurks below the surface may never be completely known. What is certain is that the future will continue to release time bombs as we begin to discover the depths and breadths of the ways in which the earth has been poisoned. A cursory examination of maps showing the distribution of cancer clearly demonstrates the non-random nature of the distribution of various forms of the disease. The factors that underlie these findings may eventually be discovered but the bottom line will likely be that the cause was the technology that is supposed to make everyone's life better.

Insidious Cumulative Effects

One of the more difficult issues in determining the effects of various chemicals is the element of time. Many environmental toxins are undetected and hidden. Their presence may not be known, nor may the associated dangers of various chemical compounds be fully recognized.

Beyond the effects that can be determined from simple exposure to certain chemicals or reactions such as radiation there are the cumulative effects of multiple toxins. Several chemical compounds such as benzene and ethylene dibromide (used in jet fuel) have known adverse reactions to exposure. The dangers of jet fuel were tagged as a result of the chemical seeping into the groundwater at Kirkland Air Force Base in Albuquerque, causing kidney and liver damage in the population around the base (PBS, 2014, May 13). The effects of most chemicals, unfortunately, because of the processes by which compounds are determined to be dangerous, are unknown.

Those chemicals that remain untested or are guarded as proprietary information, or that have not been in existence long enough for their effects to be known, are cause for great alarm. More alarming are those chemicals that are unlikely to be damaging in small amounts but produce major effects when they accumulate within the organism. Examination of blood and tissue samples by the non-profit Environmental Working Group

(EWG) found 455 industrial pollutants in the bodies of almost all Americans (Savan, 2007). Environmental toxins are so ubiquitous in the tissues of most humans that researchers have long been able to determine socioeconomic status by examining which toxins are present (Evans & Kantrowitz, 2002; Tyrell, Melzer, Henley, Galloway & Osbourne, 2013). Other studies, as noted, have detected the presence of chemicals such as flame retardants in the blood of those belonging to near–Stone Age tribes (Cribb, 2014). Alarmingly, a Belgian study found carbon particles on the fetal side of placentas (Bove et al., 2019). Exposure to these have been related to miscarriages and low birth weights.

One of the problems in attempting to delineate the effects of any relatively new phenomenon is that some effects may go unnoticed for extended periods. Several studies have found declining fertility rates as measured by sperm count among users of various technologies (Phillips, 2011). Studies, for example, have found that males who carry cell phones in their front pockets experience dramatic decreases in sperm count and motility (Agarwal et al., 2008; Wdowiak et al., 2007). Other studies, perhaps more alarming, have found a general decline in sperm counts (Vierula et al., 1996; Suominen & Vierula, 1993; Carlsen, Giwercman, Keiding, & Skakkebaek, 1992). These findings have been attributed to many empirically validated factors including fertilizers and electromagnetic waves (Agarwal & Said, 2010, Li et al., 2010). However, the primary culprits appear to be linked to increases in the use of hormones and endocrine disrupting chemicals (EDC) such as BPA in the food chain (Aksglaede, Juul, & Leffers, 2005; Den Hond & Schoeters, 2006; Diamanti-Kandarakis, Bourguignon, & Guidice, 2009; Roy, Chakraborty, &Chakraborty, 2009; Schell & Gallo, 2010; Massart, et al., 2006). Problems with male fertility have been noted to be specifically linked with endocrine disruptors and estrogens leached from plastic (Toppari et al., 1996; Skakkebaek, Rajpert De Meyts & Main, 2001; Lassen, Iwamoto, Jensen & Skakkebaek, 2015).

The overall conclusion is that things humans have engineered—i.e., technology—are the underlying causes. The effect of altered fertility is, however, not so clear-cut. In a peculiar turn, as male fertility appears to be on the decline, female reproductive periods are expanding. The age of menarche appears to be occurring at an earlier age, with a mean age of 12.34 in 2002 compared to 14.2 in 1900 (Anderson & Must, 2005; Steingraber, 2007). In addition, the age of the onset of menopause is increasing (Dratva et al., 2009).

An interesting finding related to these effects concerns adrenogenital length. The area between the genitals and the anus is substantially longer in males than women. Or at least it used to be. Evidence is emerging that adrenogenital length in males is shrinking while it is increasing in females.

An interpretation of this evidence suggests that the biological differences between the sexes are being minimized as a result of environmental factors. There is persuasive evidence to suggest that the environment is having biological consequences on the human species. The prevalence of individuals with intersex characteristics and variations has been determined to be increasing and the evidence suggests that the effects are the result of exposure to endocrine disrupting chemicals. In a study examining the parents of children with intersex variation, significant occupational exposure to endocrine disrupting chemicals was found (Rich et al., 2016). This same study noted significant intersex variations and even sex reversal in fish and reptiles exposed to EDCs. The ability of EDCs to produce epigenetic changes to adrenal functions as a result of exposure has also been empirically demonstrated (Martinez-Arguilles & Papadopoulos, 2015).

Dying for Convenience

Of all the species that are impacted by climate change there is only one that has conscious awareness of what is occurring. This species, *Homo sapiens*, is also the one most responsible for the damage and change. As previously noted, the advent of the Industrial Revolution can be seen as the demarcation point at which humans began to substantially alter their environment. The effect upon the land, climate, and other animals is noteworthy. However, from the vantage point of what humans are doing to themselves, industrialization appears to be slow suicide.

According to the World Health Organization (WHO), pollution is directly responsible for the deaths of 8,000,000 people each year. In other words, 1 in 1000 people on earth will die as a direct result of the products of technology and modern existence (UN, 2015). Toxic air alone is thought to kill 3.3 million people per year with 4000 people per day dying in China alone of pollution (Lelieveld, Evans, Fnais, Giannadaki & Pozzer, 2015). A WHO study put the numbers even higher with 3.7 million estimated to die each year from pollution; the study further found that the rates had quadrupled over the preceding 40 years (WHO, 2012). Other sources have reported 400,000 a year dying as a direct result of the effects of climate change (Friends of the Earth, 2015).

In addition to killing people outright, the by-products of the technological era have other deleterious health effects as well. By 2050 it is estimated that 50 million to 350 million people will be forced to relocate as a result of climate change from reasons ranging from rise in sea level to desertification (Whitmee et al., 2015). For a comparison, the same study found 20 million displaced in 2008. Given this year as a baseline, the cumulative numbers predicted for 2050 appear to be a severe underestimate.

Displacement and relocation are not the only noteworthy effects of the technological revolution upon the plan. In sum, climate change is just one of the more obvious symptoms. More importantly, the planet is being poisoned along with everything on it. This problem is not just a few distinct disasters and isolated incidents. Rather, what becomes clear as the evidence emerges is an interlocking, coherent picture of the industrial contamination of the biological world.

In what is unlikely to surprise, the disproportionate impact of the effect of environmental toxins is upon the poor and disenfranchised. In effect, those who benefit the least from the transition of the world from a largely pastoral or agrarian society to one guided by technology and innovation, are the most affected by its by-products. No matter whether it occurs as the result of illegal dumpsites, exploitive and environmentally irresponsible mining practices, poisoned rivers, or a host of other sources, the disproportional effects of these problems fall on the poor.

Two reports, one issued by the World Bank titled "The impact of toxic substances on the poor in developing countries" (Goldman & Tran, 2002), and the other by the Global Alliance on Health & Pollution (GAHP, 2013) titled "The poisoned poor: toxic chemicals exposures in low- and middle-income countries," detail and delineate these extreme effects. The material contained within these reports is shocking. From the proliferation of dumpsites to the presence of pollution producing industries in or near residential areas, the poor are disproportionately affected.

Table 3.2. Effects of Toxins on the Poor

Toxin/Source	Exposed Population	Nations
Lead poisoning—Battery recycling	20 million	Mexico, China
Mercury contamination— Gold mining	17 million	Congo, Brazil
Chromium contamination— Industry	15 million	
Heavy metals—Mine tailings	25 million	
Electronics waste	3 million	
Arsenic	60 million	Bangladesh
Pesticide exposure	Billions	India, Argentina, Brazil
Persistent organic pollutants	Billions	Kazakhstan, Vietnam, Africa

SOURCES: Goldman & Tran, 2002; Global Alliance on Health & Pollution, 2013.

GAHP identified more than 3000 toxic "hotspots," in its Toxic Sites Identification Program (TSIP). To demonstrate the extreme effect these toxins can have, one need only to examine the effects on IQ scores associated with lead exposure. According to landmark studies conducted across a number of countries the effects of even low levels of lead in the blood are significant, ranging from approximately 5 to 7 points (Lanphear et al., 2005; Schwartz, 1994). High blood lead levels are associated with even greater cognitive deficits, with average IQ losses ranging from 10.32 points (Lanphear et al., 2005) to 14.96 points (Schwartz, 1994).

The implications of these findings are staggering, especially given the numbers exposed. For a means to see how severe this could be in purely economic terms consider a study examining the effects of mercury pollution in the U.S. The loss of dollars as a result of lowered IQ associated with mercury toxicity was calculated to be in excess of $8.7 billion, affecting 600,000+ children in the United States (Trasande, Landrigan & Schechter, 2005).

Dollars, however, are not the most important metric by which to examine these issues. The tremendous effects upon the lives of those affected by lead poisoning and exposure to other neurotoxins are where the real damage lies. At heart, it is the stealing of futures of people and of nations. The recent discovery of lead in the water of Flint, Michigan, has made clear that this problem is not confined to developing nations. One report found nearly 1400 water systems across the country have lead in the system that exceeds safe levels (Foley & Hoyer, 2016). Another study found even greater numbers, with 3000 systems reported to have levels higher than those detected in Flint (Reuters, 2016, December 20). The result of various municipalities across the country having contaminated water supplies will take years to determine. However, it should be expected that developmental delays and cognitive deficits will occur.

These losses, in the cognitive area alone, question the whole value, economic and social, of technological existence. As these practices continue, most either try not to think about it or avoid the information at their fingertips. No one wants to think about nine-year-old miners working in death traps to dig out the rare earths needed to make their smart phone work. The issue is not just monetary, it is moral. Some products tout their production methods, with labels such as "organic" or "free trade," while other products, such as cigarettes are required to have warning labels that the products are harmful. What should the label on a computer or smart phone be—"This product may have been made with slave labor"?

Everything Gives You Cancer

With very little effort one can quickly access demographic and epidemiological information on a wide variety of topics. One of the more interesting visual representations that can be found online is that showing the incidence of cancer throughout the world. Known colloquially as "cancer maps" these graphic representations of the incidence and prevalence of cancer, whether broken down by country, state, county, ZIP code or some other means, quickly reveal an almost inescapable fact. Cancer is not randomly distributed. Certain types of cancer occur far more frequently in certain locations. What is even more noteworthy is that the relationship between cancer and the types of local industry, military installations, nuclear facilities, etc., that are present is also anything but random.

A few examples of this phenomenon demonstrate the relationship to technological development. Thyroid cancer, for example, demonstrates a clear pattern of occurrence. Increases in the incidence of pediatric thyroid cancer were observed in a direct relationship to distance from Chernobyl following the nuclear accident there (Nikiforov & Gnepp, 1994). A study examining the incidence of thyroid cancer in the United States found clearly identifiable clusters in proximity to nuclear facilities (Mangano, 2009).

In a series of longitudinal studies conducted in Finland involving examination of 15 million people across five nations, researchers were able to differentiate cancer risks based on occupation (Pukkala et al., 2009, 2014). In sum, these studies found that the exposures to toxins within various occupations are largely responsible for the types of cancers found among workers. Except for skin cancers in those who work outdoors, such as farmers, cancer variation by occupation is almost solely a function of occupational exposure to toxins. A search for "cancer" and "environmental factors" on Google Scholar will result in nearly 4 million peer reviewed articles on the subject. While it is unlikely that every single one of these finds a positive relationship of environmental toxins to cancer, the very nature of the peer review process suggests that this is a staggering number of studies connecting these two topics.

Of course, it does not have to be something as dramatic as cancer for the effects of pollution to be seen. The effects of environmental toxins, teratogens, have been previously explicated with respect to neurological disorders caused by known exposures. Evidence is beginning to emerge that other health effects are also becoming apparent as a result of an increasingly toxic environment.

Alzheimer's and Autism

The increase in the rates of autism have been staggering over the past 30 years going from a 1 in 500 phenomenon to one that is less than 1 in 100. While some of this increase is related to factors such as shifting definitions of autism and better methods of detection, the increase in rates is nevertheless truly astounding. Genetic explanations do not provide an answer either. Autism is considered to be polygenic—i.e., multiple genes are thought to be causative. However, given that persons with autism tend to reproduce in lower numbers a purely genetic transmission mechanism should then lead to it being slowly bred out of the population rather than showing relatively fast increases.

A growing body of evidence is starting to point towards environmental factors in autism. In a Danish study, a drug used to control seizures, valproic acid or valproate, was found to be associated with an increased risk of autism for offspring (Christenson et al., 2013). Other studies have found increases in use associated with maternal antidepressant use (Croen, Grether, Yoshida, Odouli, & Hendrik, 2011). Additionally, high blood levels of mercury in mothers have also been found to be associated with increases in rates of autism (DeSoto & Hitlan, 2007; Kern, Geier, Sykes, Haley & Geier, 2016). These substances are known teratogens and likely it is not a complete surprise that these seem to increase risk.

Before proceeding, a sad chapter in this search must be addressed. *Vaccines do not cause autism.* The study that originally implicated vaccines was faked. The British medical journal *The Lancet*, which published the original study, devoted an entire issue to retracting the story. The researcher who faked the results has been fired and drummed out of academia. Unfortunately, this bit of misinformation persists in large part due to a few well-meaning celebrities with no scientific training pushing anecdotal information.

A more alarming causative agent has begun to emerge as the probable culprit. This agent may be more alarming than that associated with the use of medication or exposure to neurotoxins; it is something that is everywhere and about which there are no short-term solutions: air pollution.

In a series of studies carried out in California, air pollution and its analogs have shown a strong association with autism. The CHARGE (Childhood Autism Risks for Genetics and the Environment) study found that a "minimum of 40% of autism cases are likely to have an environmental cause" (Hertz-Picciotto et al., 2006). Another study examining the concentration of air pollutants found a link between particulate matter in the air and autism (Windham et al., 2006). Ambient pollution levels and a link to autism were also found in a study conducted in Los Angeles (Becerra,

Wilhelm, Olsen, Cockburn, & Ritz, 2013). Finally, a relationship between the prevalence of autism and the proximity of the mother's residence to a freeway was found (Volk, Hertz-Picciotto, Delwiche, Lurmann, & McConnell, 2011).

A number of studies have also linked air pollution to dementia (Calderon-Garciduefias & Villarreal-Rios, 2017; Chen et al., 2017; Wilker et al., 2016) and coronary issues (Dorans et al., 2016; Rice et al., 2015). The relationship here is clear and to some degree obvious. At a very basic level the association between Alzheimer's and air pollution can be described as dirty air leads to a dirty brain. Although these studies have received some attention in the popular press there has been little attention given to the efforts needed to reverse these problems.

Connections between diseases and disorders that occur as the result of unknown and/or insidious exposures that have cumulative effects are difficult to understand on a direct, experiential level. Scientific illiteracy and industry denial hamper efforts to bring awareness of such associations. Further, as the Flint water crisis demonstrates, even when there are known risks, it does not mean that we are free of them. The lead in the water in Flint was entirely preventable yet occurred because of efforts to save money. Profits before people is the norm.

Profits Before People: The Fuels that Power Technological Society

While a devotion to technology as the only path forward can lead to an inability to see the downside of its uncritical acceptance, greed can lead to unconscionable behavior in service to profits. The history of the fossil fuel industry is one checkered with propaganda efforts, price manipulation and a disregard for basic safety procedures. In the petroleum industry, spills are common, yet predictable. In almost every case listed below, violation of basic safety procedures either caused or worsened the event. At times, events surrounding the accidents appear almost willful. The BP Gulf fiasco alone caused massive irreversible damage that forever altered the ecosystem and economics of the Gulf Region of the United States.

Table 3.3. Major Oil Spills

Responsible	Where	When	Amount (million gallons)
Gulf War	Arabian Gulf/Kuwait	Jan 1991	380–520
BP Horizon	Gulf of Mexico	April 2010	200+

Responsible	Where	When	Amount (million gallons)
Ixtoc 1	Mexico	June 1979	140
Atlantic Empress	Trinidad & Tobago	July 1979	90
Fergana Valley	Uzbekistan	1992	87.7
Kolva River	Russia	August 1983	84
Nowruz Oil Field	Iran	February 1983	80
Castillo de Bellever	South Africa	August 1983	79
Amoco Cadiz	France	March 1978	69
ABT Summer	Angola	May 1991	51–81

Note: Only oil spills of more than 50 million gallons are noted above. There have been thousands of oil spills in history. The above were also significant because of the severe environmental impact that resulted. By comparison, the Exxon-Valdez disaster would barely make the top 40 in terms of gallons of oil spilled (at an estimated 11 million gallons).

SOURCES: Mohit (2019, October 10). 11 major oil spills of the maritime world; available at Marine Insight.com; Moss, L. (2010, July 16). The 13 largest oil spills in history, Mother Nature Network at mnn.com/earth-matters/wilderness-resources/stories/the-13-largest-oil-spills-in-history; Rafferty, J.P. (n.d.). Nine of the biggest oil spills in history. At Britannica.com/list/9

The damage done just by the spills listed in Table 3.3 is incalculable. In addition to the major spills listed above are the constant small level leaks in pipelines, often running through wilderness areas, and the loss that occurs in the extraction process. Greenpeace has estimated that the Keystone XL pipeline will see 59 spills in a 50-year period if it has only an average rate of spills (Greenpeace, 2017). Such accidents will continue, and the emissions produced from the intended burning of oil will continue to increase as more people begin to drive. The obvious catch here is that as more people economically benefit from the digital economy the costs of that economy on the environment also increase. The only part of the equation that does not change is the seeming inability of industry to address safety concerns. According to a report issued by the U.S. Chemical Safety Board (2014) there were multiple failures that resulted in the BP Horizon blowout and the potential for another catastrophe remains. One of the major issues highlighted by the investigators was that of "bad management."

Coal and Natural Gas

The damage done by coal in both emissions and the process through which it is mined and readied for use is one of the more salient issues in global warming. Mining disasters are so common that it is impossible to even begin to delineate them. A study examining mining deaths in China over a 7-year period found over 9000 accidents and 23,000 deaths (Robson et al., 2011). Further, the mortality rates of miners as a result of occupational exposure are well documented (Hendryx & Ahern, 2009). Yet the need for dirty coal constantly exacerbates a problem that future generations will have to deal with. Toxic chemical spills such as the one that caused West Virginia to shut off the water supply to 300,000+ residents in early 2014 is just one of the too numerous to count releases of toxins into the natural environment that due to the constant level of leaks and accidents goes mostly unnoticed unless it impacts a major population center in a way that is inescapable. In California, the "Big Leak" spewed close to a billion gallons of natural gas from the Aliso Canyon Storage Facility. This leak was so bad that it sickened people nearby and gathered attention for how long it continued releasing nearly 10 million metric tons of CO_2 which is equivalent to burning nearly a billion gallons of gasoline (Environmental Defense Fund, 2016). Unfortunately, leaky natural gas wells are common with thousands across the nation leaking.

However, petroleum and coal are not the only raw materials of the industrial age that produce toxins. One emergent fuel source, natural gas, is heralded as something of a godsend as it removes U.S. dependence on foreign oil. However, the increase in natural gas production has been aided by the process of hydraulic fracturing, or "fracking." This approach, while essential to the production of natural gas, has produced a set of unexpected problems that must be addressed. It has been blamed for contributing to earthquakes in at least five states: Oklahoma, Texas, Ohio, Arkansas, and Colorado. Several of the states' geological services have stated that fracking is the cause of the instability despite considerable pressure to not make this pronouncement. The incidence of low-level earthquakes has risen dramatically in the past few years and directly parallels the increase in fracking activity. The United States Geological Survey (USGS) has noted that there is strong scientific consensus that earthquakes are "caused" by the disposal of wastewater produced by oil and gas exploration and extraction. These effects are especially prominent in Oklahoma (Borenstein, 2016).

In Oklahoma, for example, there were on average two earthquakes a year until the advent of fracking. In 2013 there were 109 earthquakes and in 2014 there were 585 (PBS News Hour, April 22, 2015). These quakes have been unambiguously blamed on fracking activity by the United States Geological Survey (USGS, 2015). Finally, in 2016 after earthquakes begun to occur at greater frequency and magnitude the state of Oklahoma closed a number of wells. On an intuitive level, the idea that blasting millions of gallons of pressurized fluids into the ground could create seismic instability seems obvious. Some expectation of this result should have been anticipated or at least evaluated. Fracking is a perfect example of throwing the precautionary principle out the window. Fracking may or may not ultimately be demonstrated to be a safe, environmentally sensitive procedure, but to consider that to be the default evaluation is absurd.

Other considerations it would seem prudent to address before attempting this approach would be the concern of groundwater/water table pollution, along with attendant effects on humans and wildlife. The fact that proprietarily secret chemicals are used to blast into the earth would seem to further demand that safety of the process be demonstrated.

Instead, the opposite is found. The companies trumpet their methods as "safe" and manipulate the public discussion. It is not that the materials and procedures are safe. Rather it is that the appropriate studies have not been completed. Further, reluctance and outright resistance exist to having the studies conducted by outside academics and therefore impartial researchers. The reason there are so few studies is more a function of barriers to the studies and the extremely slow nature of the peer review process than of a lack of evidence. Further, any study that finds problems will likely have to confront a strategy of "reformulation" of the secret formula.

Despite the significant barriers to conducting scientific studies there is evidence that the fracking process also contaminates ground water and affects air quality. A study conducted by the Colorado School of Public Health and Brown University found that residents living within a half mile of wells experienced significant and increased health risks including higher rates of cancer and neurological deficits (McKenzie, Witter, Newman & Adgate, 2012). Alarmingly, an association with birth defects was also suggested by the research with an increased risk related to maternal residence within 10 miles of a well (McKenzie et al. 2012). Unfortunately, there is more to worry about.

The ultimate expression of the technological age as regards fuel is nuclear power. Presented as an almost magical solution to the rising needs for energy, it is hailed as clean, safe and cheap. It is none of these things. A more extensive discussion of this topic will follow in later sections of this book.

The Biotic Right

Arguing for the rights of the *biota*, the plant and animal life, is likely to be categorized by some as an esoteric, or even flaky concern. However, the rights of the biota, while it is possible to incorporate this view into a Gaia or stewardship-type religious framework, will be considered herein from the point of view of reason and survival. In essence, the biotic right is also a human right in that humans are part of the biota. A critical feature missed by those who shrug off the destruction of the natural world as a necessary feature of "progress" is the inescapable fact that every single input into the physical production of every piece of technology originates in the natural world. When the natural world is destroyed to the degree that it is uninhabitable for humans, humans will cease to exist.

When framed that the survival of humanity is about the preservation of the natural world, the implications are clear. The preservation of nature is far more important in the most basic sense than the promulgation of technology. Sustainability is not some leftist conspiracy as envisioned by corporate America; it is the only logical approach to the use of finite resources that must theoretically last the planet for infinity. Unfortunately, short life spans and poor ability to envision the future prevent humans from thinking in terms of infinity. As a result, policy tends to be made only a few years into the future and anything attempting to plan even 50 years into the future is unheard of.

Beyond short-sighted thinking that seems to deny the obvious connection between technological development and the destruction of the natural world lie other barriers to the creation of anything resembling a sustainable approach to technological growth. Primary among these barriers is an alienation from nature that leaves many with a view of nature as hostile and threatening. "Bios-Fear," or the fear of nature, a punning term coined by novelist Ray Bawarchi (2007), aptly describes this approach. Phrases such as "battling the elements," "conquering a mountain," "struggling against nature" all pay homage to this mindset.

At a very deep level, alienation from nature is alienation from oneself. Humans are organic beings. That seems an obvious point, yet the noted desire to merge with machines belies acceptance of this fact. Underlying the urge to become one with the machines is an unspoken and perhaps unrealized belief that the machines are superior. The whole extropian movement and all efforts to enhance brain functions are premised on the idea that being an organic being is somehow less than being a machine. Arguments that humans must be enhanced in order to keep up (e.g., Kurzweil, FM 2030) are unwittingly diminishing the rights of organic beings.

At a fundamental level all arguments of resource use boil down to whether one thinks the items produced by the resources are more valuable than the resources in their natural state. The other factor in this equation, the one conveniently overlooked by polluters, are the costs of the use of the resources. When this factor is taken into account the destruction of the natural world suddenly becomes a no-brainer. The consequences of the current course of action are too great. A different direction must be determined.

A fundamental shift in the manner in which the destruction of nature is viewed is required. Specifically, the inherent violence of technological development as regards the plunder of the natural world must be acknowledged. There is no production of computers, robots, smart phones, automobiles, planes, etc., without a direct impact on the biological world. At the current pace of destruction, humanity is headed towards a point where most people will live in large cities and nature will become relegated to museum-like status. Just as zoos are the final place for many threatened species, nature preserves may become the norm for the organic world.

But zoos and nature preserves are just illusions. They serve to create a false sense that hides the horrible realities. Vast stretches of the natural world are quickly being destroyed, all in service to an artificial notion of progress that equates technology with value. One of the more stark examples of the effort to promote technology over the natural world is an effort currently underway at the Smithsonian Biology Institute in Virginia. They are attempting to protect threatened and endangered animals though the use of reproductive technologies such as in vitro fertilization (Akpan, 2015). The idea that the solution to the looming extinction of many animals is to aid their reproduction through artificial means borders on asinine. The problems that endangered animals have are not an inability to reproduce but rather a lack of a place to reproduce. Nature relegated to labs and museums is not nature. It is surrender, but it calms us into thinking that everything will be all right because technology can reverse the damage.

The acceptance of a machine-oriented existence may be presented as the way of the future. Suggestions that the only way forward for humanity is a technologically based existence ignores the reality of choice. There is a trade-off between humanity and technology that at present is only beginning to enter human awareness. Decisions must be made as to what is more valuable: life or gadgets?

The right of the natural world to exist must be acknowledged. What seems to be a fundamentally evident point, that the planet and its ecosystem is more valuable than the development of devices that do not currently exist, appears to require restating. The destruction of the natural world and the killing of the environment, ecocide, cannot be casually accepted. The right of the biota, down to the lowest single-celled organism, must be seen as having primacy over the rapacious consumption necessary to continue the technological age. Is this likely to happen even though it may be necessary for the survival of humanity?

4

The Coming Cataclysm

A Runaway Train Jumps the Tracks

There is no Planet B.—Environmental meme

By the time it was noticed that the moths in England were evolving in coloration as a result of the soot pumped out by early industry, human activity had already begun to slowly increase the carbon levels in the atmosphere. That effect has continued unabated and its cumulative effects are becoming obvious.

Beginning in 2011, the Intergovernmental Panel on Climate Change (IPCC) began issuing comprehensive reports that attempted to assess the effects of climate change. Issued in the reserved style of academia, the reports are noteworthy not only for their presentation of vast amounts of data related to climate change but also for their increasing efforts to raise awareness in the public.

The fourth volume of the report issued in 2014, was unequivocal and alarming. Using terms such as "irreversible" and "catastrophic" to describe the future the report is distressing (Ritter, 2014). With each passing year the effects get worse and the warnings more severe. The IPCC became unequivocal in its statements conclusively putting the blame on human activity for climate change and implicating climate change in the increasing volatility of weather events (IPCC, 2014–2019). The panel, again comprised of scientists who are normally understated in rhetoric, found the problems so severe that it advocated efforts it had previously eschewed, such as attempts to modify the weather, to address the problem (IPCC, 2018). At this point, it appears that even the scientists trying to raise the alarm have given up.

With near scientific certainty (IPCC, 2015), the burning of fossil fuels to power the various machines necessary for the current era is causing the climate of the earth to change. From 1976 until the present EVERY year has exceeded the mean annual temperature for the planet for the past 11,000 years (Marcott, Clark, Shulkin & Mix, 2013). At a very basic level this means that no one who was born in the last four decades has a good idea what the climate should be like.

There has been an average increase of 0.50° F per decade over land and 0.20 degrees over the oceans (NOAA, 2019). Since 1900, the average global temperature has increased by 1.8° F and sea levels have risen by 7.8 inches (Wuebbels, Fahey & Hibbard, 2017). As the earth gets warmer, sea levels rise, coast lines shift, and the weather becomes more extreme. Effects are not limited to matters as simple as these. In addition, as the weather shifts and the planet heats up there will be environmental degradation, habitat loss and challenges to numerous species.

In an alarming 8000-page report, the Intergovernmental Science-Policy Platform on Biodiversity and Ecosystem Services (IPBES, 2019) released findings summarizing 50 years of studies on the state of the environment. The most alarming information in the report details the absolute havoc being wrecked upon nature. The biodiversity of the planet is imperiled, and more than one million species are at risk of extinction (IPBES, 2019). This report, coupled with the IPCC reports of the past decade, demonstrates a clear scientific consensus. The world is in trouble. The activity of the human species is leading to the destruction of the natural world and by extension, the human occupants.

By every metric, the planet is in trouble. We have reached a state where the natural world is on the brink of extinction. Biodiversity is rapidly diminishing and the habitat of the few animals left is shrinking. The climate has turned against us and the seasons are beginning to fluctuate.

It is estimated that an acre of rainforest disappears every second, amounting to 85,000 acres a day with 1.5 million acres lost in just the Amazon rainforest in 2014 alone (Rainforest Alliance, 2015). The loss of just primary forest (i.e., old growth, naturally occurring) since 2000 is estimated to be 2.3 million square kilometers (Whitmee et al., 2015). The average rate of loss of forest is estimated to be approximately 14.5 million hectares (1 hectare = 2.7 acres) per year (UN: Global Forest Resource Assessment, 2010). South America, primarily as a result of logging and corporate monoculture, has experienced losses of nearly 30 percent of its wilderness, mostly rainforest, in the past 20 years (Watson et al., 2016).

Trees are the metaphorical lungs of the planet. Not only does deforestation affect habitat it also alters the ability of the planet to store carbon and clean the air. However, deforestation is not the only means by which habitat is lost. Entire ecosystems and the biodiversity within them are disappearing. Precipitous declines have been noted in the last remaining areas of wilderness. Approximately 10 percent of all wilderness areas have been lost in the last 20 years, with South America (30 percent) and Africa (14 percent) experiencing the greatest losses (Watson et al., 2016). Examination of 13 different categories of wilderness ecosystems found all had lost habitat and only two (tundra and boreal forests) had more areas designated for protection than were destroyed. For all except two categories most (> 50 percent) of the original range is gone. A stunning fact somewhat obscured by the data is that almost all wilderness left is in areas completely hostile to human habitation (Watson et al., 2016).

The planet may have reached carrying capacity. With almost all wild areas restricted to regions too hostile for human habitation, nature is essentially dead. Habitat destruction will assuredly continue as well as the demand for more resources for an exponentially growing population. The inability to replace what is being taken can only end in predictable catastrophe.

Water: The Essential Element of Life

The water systems of the planet are also in trouble. Only 13 percent of the world's ocean are free of human pressure (Jones et al., 2018). These few areas, again similar to wilderness areas above, appear to be fortunate because they are almost entirely in areas of low human population and devoid of natural resources.

The water of the earth exists in a complex dynamic system. In elementary schools, children are taught about the water cycle as part of the science curriculum. Without water, life on the planet would cease to exist. Water maintains life, water is life. Yet an outside observer would hardly conclude that humans appropriately value water, given the way it is treated. By nearly every measure the water supplies of earth are disappearing or being degraded.

One important component of water on earth are the wetlands. They serve to filter toxins and act as crucial habitat for many species. Over 100 million acres of wetlands are estimated to have been lost in the United States between the time of European settlement and 1980 (Dahl, 1990). In Illinois alone it has been estimated that 23 percent (3.2 million hectares) of the state was covered in wetlands prior to westward expansion but that 90 percent of that has been lost to urban development and agricultural use over time (McCauley & Jenkins, 2005). Since that time the problem has accelerated. Fifty percent of salt marshes in coastal areas around the world are believed to have disappeared (Barbier et al., 2011). On the west coast of the U.S. this problem is even greater with an estimated 90 percent of these resources destroyed (Gibson et al., 2015; Barbier et al., 2011). Global warming will further exacerbate this issue, with nearly 70 percent of all coastal wetlands already endangered by rising sea level (Blankespoor et al., 2014).

Beyond wetlands and forests, other crucial habitats are also threatened. From the prairies to the ice fields, habitat loss is occurring faster than the ability of animals to adapt. Polar bears are struggling to find hunting grounds, sharks are struggling to find food and many other animals are being squeezed out of their habitats and into extinction with little to no fanfare. The alienation from nature bred by a technologically focused culture render these effects largely out of sight and out of mind for the vast majority.

Most, if not almost all, of the damage inflicted upon the natural world by technological development has been unintentional and accidental. While the intent has not been to deliberately inflict damage, severe degradation of the environment has been a necessary consequence, and oft acknowledged outcome, of that which is deemed to be progress. Damage is rationalized as for the greater good.

This tradeoff between environmental damage and that viewed as technological progress is rarely subjected to a legitimate cost-benefit analysis. The assumption is that progress must occur. In the not too distant past, there was little thought given to such notions as resource depletion. For much of the early part of the Industrial Revolution, which coincided with a period of imperialistic expansion and attendant colonialism, resources were plentiful and environmental damage went largely unnoticed. By the time science began to even suspect that human activities were negatively impacting the environment, both a consumer mentality and the deification of technology had occurred. Concern for the effects of one's actions, the precautionary principle, or an environmental ethos had yet to be envisioned. The genie was out of the bottle and didn't know its own strength.

Unnoticed and unseen the genie moved in slow motion. The effects of environmental degradation were masked for centuries due to the relatively slow pace at which it proceeded. An accumulation of effects and an acceleration of activity have forced awareness of the connection despite industry propagandists who attempt to inject doubt into scientific consensus. These efforts have been well documented by various media sources and scientific groups, e.g., Union of Concerned Scientists (Catalyst, Summer 2015). Just as is true with people, it is the cumulative exposure to toxins that seems to most affect nature. Slow and insidious damage can become irreversible.

While government inaction has become status quo and private industry disinformation is better funded than ever, it has become apparent that human activity concomitant with technological development has produced dire consequences for the planet. By almost any metric the effects of the development and production of technology upon the natural environment are enormous. From rampant deforestation and increasing desertification to the loss of critical habitat and the escalating die-off of species, the effects of climate change that directly result from actions required to sustain the digital era are nearly incomprehensible. The problem is so far-reaching that any discussion of the ongoing extinction of the flora and fauna of the earth's biosphere can be little more than cursory given the absolute magnitude of the problem. Despite how severe the effects are already, the one inescapable conclusion of the evidence is that the problem will continue to worsen. The effects are no longer moving in slow motion.

Rising Tides

According to a study by geophysicists at Harvard, sea level rise appears to be accelerating at an even greater pace, 2.5 times, than previously detected (Hay, Morrow, Kopp & Mitrovica, 2015). Alarmingly this study also found that previous estimates were inaccurate in a way that tended to lead to underestimates in the effects of sea level rise. Shrinking glaciers, especially in Antarctica and Greenland, are one of the major contributors to sea level rise. Entire sections of the Larsen Ice Shelf in Antarctica are gone. Places with significant areas of glaciations have changed so dramatically that satellite images no longer match maps made even a decade ago.

The effects of massive amounts of ice water being released into the ocean are likely to have even greater effects than the mere measurement of sea level. The chemical composition of the ocean is being affected in such a way that it is becoming more acidic. According to Global Oceans Commission (GOC, 2014), the acidity rate is 30 percent higher than 200 years ago and is dramatically accelerating (GOC, 2015). Natural changes in acidity and other chemical compositions that formerly took thousands of years are now occurring in a few decades. These changes are having dramatic effects upon sea life.

Most of the carbon on earth is trapped in the oceans (90 percent by some estimates). As more enters through the melting of glaciers and continued pollution by fossil fuels the CO_2 in the water begins to deplete calcium carbonate. Calcium carbonate is essential to produce shells in a wide range of species. Too much acidity causes the shells to grow slower and thinner. Increasing acidity is impacting the ability of shellfish to form their shells. This affects not only growth but also fertility rates in species such as abalone. Other animals throughout the food change are also being impacted by both pollution and the direct and indirect effects of climate change. Examples of other changes caused by global warming include colder water and lower salinity in the areas surrounding the Polar regions. As the icecaps and permafrost begin to melt the increasing amount of cold water not only affects the oceans but also affects the atmosphere through the release of carbon, methane and other substances trapped in the ice. According to the EPA, even under the best-case scenario, significant amounts of greenhouse gases will be released no matter what our responses.

Coral reefs are also affected by changes in ocean acidity. Again, as with shellfish, the acidity affects the ability of the reefs to utilize calcium carbonate. Around the world coral reefs are showing damage and dying. This matters a great deal because even though coral reefs occupy less than 1 percent of the oceans, they are home to over 25 percent of the ocean's species. The International Coral Reef Symposium of 2016 convened with the specific goal of developing a plan to tackle this mammoth problem with a quarter of all reefs having disappeared in the last few years and the health of nearly every reef on the planet severely impacted. The problem is so dire that it was estimated that no matter what was done, 90 percent of reefs would die off by 2050 (Jones, 2016). This prediction, which seemed so extreme that it was viewed as hyperbole, is already seen as optimistic.

Desalination projects around the world are a relatively new factor in the destruction of the oceans. Numerous projects exist in places such as Israel, Saudi Arabia and Carlsbad. The effects on sea life are poorly understood at present. However, the process of desalination involves the filtering of ocean water with the by-products of the process typically dumped back into the ocean. This process will substantially alter the chemical composition of the water in the local area. However, the effects will not immediately be apparent, and it will only be over time, when it is likely too late, that the unintended consequences become known.

Aquifers and Groundwater

The source of much of the drinkable water utilized by humans, aquifers are in trouble. According to research conducted by NASA's Jet Propulsion Laboratory aquifers are threatened throughout the world. Of the world's 37 major aquifers, 21 are losing more water than is being replaced. Of these 21 most threatened, 13 are considered to be beyond the tipping points for sustainability (Famiglietti et al., 2015). The lead author of the study, Jay Famiglietti, has attributed the blame for this predicament squarely on the shoulders of the agribusiness industry (PBS, 2015 June 17). Actions such as the cultivation of rice in dry climates or water intensive crops in drought-stricken areas, such as almonds in California, only serve to further accentuate the problem.

In addition to aquifers, other sources of clean potable water are also being threatened. For example, many of the world's large freshwater lakes are disappearing. The Aral Sea on the Uzbek and Kazak border has basically ceased to exist. Lake Chad is now little more than a mud puddle. The

Colorado River no longer reaches the ocean, drying up somewhere in Arizona as swimming pools fill the landscape and drain away the water (Mother Nature Network, 2015). Contentious water battles are beginning, and the future of water would appear to be one destined for increased legal contention as the population continues to grow and the water continues to disappear.

It is ironic that as the freshwater lakes, aquifers and others forms of groundwater begin to evaporate, the glaciers are beginning to melt. The release of this water, generally far from human habitation, is having the opposite effect of that on land. Water, lots of it, is being released into the ocean from the land masses where it has been locked. The problem is not just the amount of water it is also the temperature.

Glaciers

The one source of fresh water that is somewhat immune to contamination by various pollutants is that which is locked up as ice in glaciers. Unfortunately, the fact that pollutants find it more difficult to infiltrate ice does not mean all is well. The glaciers are being affected by climate change in a manner than is likely to prove far more deleterious than that of direct pollution. As the climate changes, the melting of the ice has accelerated. Researchers formerly landed on the glaciers and began to take samples. Now they have to hike, sometimes for days, to merely get to the ice. As the ice melts it not only affects the land mass and its ecosystem, the water entering the ocean affects salinity and further accelerates the rise of sea level (glaciers melting on land do affect ocean levels as the water enters the system). This water also affects the temperature of the ocean as it enters it and releases carbon dioxide stored in the glacier, further accelerating the process. At this point, it may be that the melting has become irreversible.

A more alarming scenario related to the melting of the icecaps is the effect it might have on the mid–Atlantic Current. According to the Potsdam Institute for Climate Impact Research, it is possible that the current, a conveyer like system that causes a flow of water around the Atlantic Basin, could stop or slow (Rahmstorf, 2002). The effects of such a change could disrupt the weather in a profound manner. Super storms, once the province only of movies such as *The Day After Tomorrow* (2004), could conceivably become not only realistic, but regularly occurring. El Niño, the ocean phenomenon that modulates the intensity of storms in the Pacific has deviated significantly in the recent past due to changes in ocean temperature that affect this comparable conveyer system. What was once an infrequent phenomenon is becoming normative.

Resistance to the ocean is impossible. As "tsunami-proof" walls obliterated during the Fukushima disaster demonstrated, the power of the ocean can be overwhelming. Human engineering appears incapable of building any structure completely resistant to this force of nature.

Engineering prowess aside, the biggest problem is not human ability to cope with climate change, it is the fact that climate change appears to be worsening. Industrial society had been pumping out all manner of emissions and creating lethal reservoirs of toxic substances for generations before anyone even began to notice. The environmental movement, of which the beginning might be traced to Rachel Carson's *Silent Spring* published in 1962, or perhaps to the transcendentalist of the 1800s (who mostly championed environmentalism as an aesthetic movement), is by any accounts, relatively new. The application of scientific principles to evaluate the effects of pollution on human and animal health, climate, biodiversity, and other domains is still in its infancy.

The nature by which the scientific process advances has been noted to be slow and deliberate. Scientists also talk largely in probabilities and view terms like "proof" and "certainty" as poor thinking and the mark of an amateur. Unfortunately, propagandists latch onto these same terms for their own purposes, distorting the public debate and creating uncertainty about settled scientific matters. It may be called the "Theory of Relativity," but it is not as if scientists think it is wrong. Astronauts bet their lives on it. Likewise, theories of climate change are not called "theories" because they are some crackpot idea, rather, "theory" as appropriately used in science, is the term applied to an idea believed to be the best representation or explanation of a concept, observation, or structure. When a politician or oil company spokesperson (sometimes the same person) states something to the effect that "climate change and global warming are just theories," in an effort to cast doubt on scientific findings, they are either being deliberately deceptive or woefully ignorant. If they are the former, they clearly hope their audience is the latter.

Beyond the false debate created to thwart solutions and manipulate the political apparatus there remains the problem to that political apparatus to address the problem. Even if all the world governments were to suddenly agree to an approach to fight global warming and attendant climate change, there remains the question of whether that is still possible. According to the United Nations Intergovernmental Panel on Climate Change, the level of carbon emissions already in the air coupled with the ongoing increase in the output of greenhouse gases, is so great that any opportunity of reversing the damage, or even limiting the damage to something manageable, may have already passed (IPCC, 2014; IPCC 2015). In mid–May 2014, two independent research teams found that the ice shelves in Antarctica were "irreversibly melting" (McGrath et al., 2014; Scambos, Raup & Bohlander, 2012). In other words, it is only going to get worse.

Indirect causes of changes in the ocean have been somewhat difficult to quantify but other evidence suggests that the problems of the ocean are almost certainly caused by the direct effects of pollution. The massive gyre of plastic in the Pacific is but one of these pieces of evidence. The sheer volume of plastic that ends up in the oceans is staggering. According to estimates, over 8.8 million tons of plastic makes its way into the oceans each year. The United States was the 20th largest contributor with 80,000 tons produced while China led the way at 2.4 million tons, with the island nation of Indonesia second at 900,000 tons a year (Royte & Barker, 2015). This massive dumping of plastics should be expected to continue as little to no governmental action has been developed to address this problem. The fact that this is happening in the ocean far from sight also works against finding a solution. No one country will assume responsibility and the vortices that lie where the material accumulates are essentially outside any governmental jurisdiction.

Unfortunately, even when environmental crises get sustained attention there appears to be little that can or will be done. Certain environmental disasters such as the *BP Horizon* blow-out, the wreck of the *Exxon Valdez*, or even the Fukushima nuclear breech, commanded great attention, at the time. The *BP Horizon* debacle, which killed 11 workers, demonstrates how difficult it is to determine the full effects of such disasters. It is nearly impossible to estimate how things "would have been" and make an inference about the results of such events. However, it is abundantly clear that such events are of major impact on the oceans.

According to research investigations, 1200–1300 bottlenose dolphins died as a direct result of the *BP Horizon* disaster (NOAA, 2015). The entire Gulf fisheries industry was severely impacted, and litigation will continue for decades over the effects. The true impact of the event will likely never be known. Unfortunately, large events that get press coverage are not the sum total of the problem. Rather, the constant and unrelenting contamination of the oceans by various sources tells the real tale. It is figuratively death by a thousand cuts. From the emissions of ocean-going vessels, to the release of carbon from glaciers, from the run-off from pesticides and fertilizers, to the casual and unrelenting industrial disasters pumping every imaginable toxin into the water, the oceans are being destroyed. Attitudes such as those surrounding the continued release of radiation into the ocean from the Fukushima plant—a directly stated, "Don't worry, the ocean is a big place and can handle it," ignore the stark realities. Even as fish continue to test with high levels of radiation years after the event and even as massive amounts of debris wash around the world, concern is minimized, especially by those who are responsible.

At some level, the concern for the oceans may be a case of too little, too late. According to a study conducted by the Global Oceans Commission, the world's oceans are on course to be completely fished out by 2048 (GOC, 2015). A similar study reached the same conclusion stating a "100 percent" likelihood of this effect by 2049 (Worm et al., 2006). In short, the efficiency of commercial fishing combined with the depletion of ocean stocks as a result of environmental change has created conditions that are leading to an essentially dead ocean. Massive algae blooms around the world, primarily a result of fertilizer run-off are another bit of the puzzle, starving the water of oxygen.

Even if humans manage to get some handle on the factors resulting in climate change in the near future and further manage to mount an internationally agreed upon and well-coordinated strategy that completely changes or even halts climate change, it is still going to get a lot worse before it gets better. Given the obstacles to developing a sufficient response, the pace of climate change is likely to accelerate instead of decrease. The train has metaphorically crested the mountain with a little too much speed and far too much weight.

When a runaway train finally jumps the rails, the damage can be unprecedented and unpredictable. The slow submergence of large portions of the land mass might then prove to be the least of our problems.

Climate and weather, it is often stated, are not the same. One of the more frustrating elements of confronting climate change are the deniers who use every seemingly contradictory weather event to suggest global warming and climate change are a hoax. Every cool day in the summer, or for that matter, every cold day in the winter, is used by climate deniers as "proof" of their position. Of course, blistering heat waves and t-shirt weather on New Year's Day are ignored. The World Meteorological Organization (WMO, 2020) reported record temperatures for Antarctica, while NOAA has tracked a precipitous rise in temperatures in the Arctic for several years (NOAA; 2018, 2019). In both regions, temperatures have risen to levels that are astounding.

Again, despite the deniers of climate change, the evidence is overwhelming. The climate is changing, with a general worldwide warming trend. Climate is a long-term phenomenon that is influenced by a planetary system. Weather is what happens in the short term—i.e., today. Weather is influenced by conditions in the microsystem surrounding a local area. Climate is more predictable than weather, which is susceptible to rapidly changing conditions whereas the climate is controlled on a much larger scale by variables that are more systemic.

Much of the equivocation surrounding the connection between the climate and the weather ended with the release of a special edition of the *Bulletin of the American Meteorological Society* (BAMS: 2014). The results of

20+ peer reviewed studies conclusively pointed to the connection between climate change and extreme weather events (Borenstein, 2014; BAMS, 2014). Importantly, the studies collectively suggest that the possibility of severe and extreme weather events, such as droughts, heat waves, and unusual precipitation, will increase as a result of continued climate change.

Basic science tells us that as the ice melts and the temperature rises, more water will be evaporated into the air. This water will then be returned to the earth in the form of precipitation. This basic process will result in more water and more floods. This is already happening. Flood cycles are changing. What were formerly 100-year, or 500-year floods now occur every few years. These and multiple other weather events are becoming more extreme and there is scientific consensus that human activity is responsible. While it would not be valid to describe any one storm as "caused" by climate change, the idea that the overall pattern of weather will become more volatile is given great credence (IPCC, 2013; 2014). Super storms, hurricanes and other weather events of great magnitude are expected to become more common.

Conversely, as some areas of the globe receive more precipitation, others will receive less. Droughts will lead to more wildfires and wildfires will, and have, become commonplace. Research indicates a direct correlation between climate change and the frequency of wildfires (Jolly et al., 2015; Scasta & Stambaugh, 2016; Wuebbles, Fahey & Hibbard, 2017).

Not only are humans raising the temperature of the planet, evidence is beginning to emerge that suggests we are altering the seasons. Studies examining long range climate data are finding that the troposphere is heating more rapidly in the summer and cooling more slowly in the winter, resulting in shifting of the seasonal cycles (Randel, 2018; Santer et al., 2018). Places with mild winters are slowly losing the weather associated with the season while places that depend on the winter snow melt are beginning to experience water shortages as the snow fails to fall. The greenhouse effect is no longer theoretical, it is apparent.

The greenhouse effect was first described in 1859 by John Tyndall (Hume, 2009). In this early work Tyndall established that gases absorb more energy when heat is passed through them (Tyndall, 1859). Tyndall's work and the implications thereof were not fully appreciated at the time. It was only when the climate began to noticeably warm in the late 20th century that the greenhouse effect was applied to explain planetary warming. Since that time however, an enormous amount of scientific evidence has begun to accumulate that demonstrates the severe effects of greenhouse gases.

Carbon dioxide, CO_2, is the most common of what are colloquially called "greenhouse gases." Others include methane, CH_4, and nitrous oxide,

N_2O. These compounds trap heat in the atmosphere and as such are a primary culprit in global warming. Levels of CO_2 have been tracked by the National Oceanic and Atmospheric Administration (NOAA) since 1980. In addition to current tracking measures, scientific analysis is able to determine the CO_2 levels in the past by looking at core ice samples. These results are alarming.

According to data gathered by NOAA, the worldwide average for CO_2 concentrations surpassed 400 parts per million in March 2015 (NOAA, 2015; World Meteorological Association, 2016). This level had not been measured for two million years in the geologic record. The pace of the change is accelerating and there appears almost no chance that the activities leading to this will cease or even slow. In fact, the change is definitely headed in the wrong direction. By 2019 CO_2 levels surpassed 415.19 ppm which was the highest level measured in 3 million years (Scripps Institute of Oceanography, 2019). Since tracking began, CO_2 concentrations are up nearly 20 percent, increasing by 61 parts per million in a 35-year period compared to an increase of 80 parts per million over a 6000-year period prior to human industrial activity (Borenstein, 2015, NOAA, 2015).

Reasons behind this increase in CO_2 are two-fold. The direct and easily attributable part of the increase is directly due to the emissions released by automobiles, power plants, and other industries powered by fossil fuels. An indirect and more difficult to quantify element in the increase of CO_2 and other greenhouse gases lies in the melting of glaciers. As these large blocks of ice melt, the CO_2 trapped within is also released.

The effect this has upon temperature is the primary mechanism driving climate change. It is not merely that there is change in the temperature that is leading to the drastic effects that are being seen, but rather it is the magnitude of the change that most alarming. In a landmark study examining global temperatures over the past 11,000 years it was established that over the entire time frame, with the exception of the last 100 years, the total variance of global temperature was approximately 1° C (Marcott, Shakun, Clark & Mix, 2013). However, for the past 100 years the average global temperature has been 1° C above the range of observed figures for the previous 11,000 years. In absolute terms 1° C does not seem like very much until put into perspective. Altering the temperature of an entire planet in such a short period time in such a manner is on an order of magnitude that is difficult to express. An apt comparison might be that if through the history of baseball, the batting average for all the players had been .250 and has a range of .175 to .400. Then in a one-year period, the entire league suddenly hit for an average of .425. There would surely be an investigation by MLB under these circumstances.

The National Oceanic and Atmospheric Association (NOAA) has

reported that this warming trend is accelerating as marked by the accumulation of greenhouse gases and the melting of ice locked away in glaciers (NOAA, 2014). The National Climate Assessment (NCA) has essentially come to the same conclusion (Melillo, Richmond & Yohe, 2014; NCA, 2014). Nearly every month measured for the past few years has broken records for temperature (NOAA, 2018; 2019). Not only does this trend appear to be accelerating it is a given that it will continue as reversal would require a complete change of human activity. Australia found it necessary to add two additional colors, purple and magenta, to its heat maps after the existing boundary of 50° C (122°) was no longer sufficient to describe the weather. From 1880 to 2018, the 20 hottest years on record were all from 1998 to 2018. The five hottest years were the last five and preliminary data from 2019 indicates it will join the list (NOAA, 2109).

All these things are happening and no amount of sponsored "research" by the petroleum industry in service to the propaganda of climate change denial can change these facts. According to the Intergovernmental Panel on Climate Change (IPCC), featuring a collection of international experts from around the world, it is "extremely likely," affixing a probability of at least 95 percent, that humans are the cause of global warming (IPCC, 2013). The severity of these changes is further highlighted in report after report issued by separate groups of prominent scientists. A panel appointed by the National Research Council warned that a multitude of drastic and dramatic effects are likely to ensue from continued irreversible changes caused by global warming (Gillis, 2013; NRC, 2013), while a United Nations panel has continually warned that continuing the current course of burning fossil fuels will have catastrophic effects (IPCC, 2014–2019). Among the possible problems include massive dead zones in the world's oceans, collapse of ecosystems, changes to weather patterns and even mass extinctions (IPCC, 2014; NRC, 2013: IPBES, 2019). An excellent visual presentation of the multitude of findings by the IPCC can be found in *Dire Predictions* (Mann & Kump, 2015).

The effects of altering the seasons goes far beyond simply changing the temperature. The water cycle is disrupted, the breeding range and cycles of various animals are altered, more pollen is released to compensate for degrading conditions, plants begin to bloom at earlier periods in the year. Humans have not only altered the temperature; we have altered nature.

Climate change is starting to affect the world geopolitical situation. According to the IPCC (1997, 2001) the Maldives, an island nation in the Pacific, will likely be completely submerged by 2100 (United Nations, 2014). With 325,000 people living on the island group, located hundreds of miles from the nearest land, rising tides are not just a matter of lines on a map. At only one meter above sea level, the Maldives are the world's lowest lying nation. Their struggle to relocate and deal with the rising ocean is documented in "The Island President" (2011).

In Myanmar, rising tides are already having an adverse effect on the Rohingya, an ethnic group from Bangladesh. The Rohingya, a predominantly Muslim minority, face a great deal of animosity and discrimination in the largely Buddhist nation. The plight of this group appears intractable as there is simply no place for them to go. They have not only been refused entry into Indonesia, Malaysia and Thailand; they have even been towed out to sea by the naval forces of these nations (Bawarchi, 2012; Reuters, 2012). The land of their origin, Bangladesh, is disappearing at a rate of one meter per year and rising seas are expected to claim at least 20 percent of the nation by the most optimistic predictions by the year 2050, while other estimates put the country completely underwater by then (GermanWatch, 2004; Bawarchi, 2012; Lynas 2008).

Most of the world's population lives on or near a coastline. The United Nations Intergovernmental Panel on Climate Change (IPCC) estimated that sea levels could rise over 20 feet if the Greenland and Antarctica ice sheets were to completely melt (IPCC, 2014). At that point the problem begins to affect more than a few low-lying islands or defenseless ethnic minorities. At levels of a 20 foot rise, social upheaval would be almost assured. Unfortunately, this same report indicated that the earth may have already reached the point that this is inevitable and irreversible. The Greenland ice mass loss has been accelerating, experiencing a four-fold increase from 2003 to 2013, and appears to be "breaking up" (Bevis et al., 2019). Additionally, the ice mass in Antarctica, which is even greater, appears under complete retreat, with NASA using terms such as "irreversible" to describe its collapse (Hanna, 2014). Glacial events are happening at great speed.

Resource Depletion: Planned Obsolescence and Capitalism, or Follow the Money

Only after the last tree has been cut down and the last river poisoned, and the last fish caught will you realize that you can't eat money.
—Cree Prophecy

The United Nations has advocated that global emissions resulting from the burning of fossil fuels will have to drop to zero by the end of the century or temperatures will rise to levels that will result in an irreversible calamity (Ritter, 2014). The probability that such a drop will occur is the same as the levels to which emissions will need to drop—zero. Since their initial plea, matters have only gotten worse, with the United Nations now stating that reduction in emissions must occur at greater than twice the pace (7.6 vs. 3.3 percent) per year to achieve the goal. Further, they note that the current trajectory will result in a warming of more than 7° F by 2100 (UN Global Emissions Report, 2019).

The problem in a word is capitalism. The profit motive is fundamentally at odds with any efforts to control or limit emissions. In an excellent treatise on the subject, *This Changes Everything: Capitalism vs. the Climate* (2014), author Naomi Klein superbly makes the case that the confluence of blind allegiance to the free market and an expanding economy render efforts to combat climate change nearly impossible. The focus on growth as an economic model complicates all efforts to reduce consumption. As billions of people throughout the world clamor for the goods associated with economic prosperity, capitalism is not likely to withdraw to the sidelines. Sustainability along with efforts to limit pollution and other forms of environmental damage are antithetical to the free market.

The free market/capitalism is geared only towards the production of profits. Everything else is secondary. Any means necessary to increase profits is the driving force of the global economy. One practice—planned obsolescence—is at the cornerstone of profits and one of the greatest threats to the world's resources.

Planned obsolescence is the process by which corporations produce inferior products design to wear out or be supplanted by a newer model, thus guaranteeing a steady market. In the digital world the process is disguised by the release of "upgrades" or "next generation" products.

Producing a product that will need to be replaced is clearly good for the profit motive. Conversely, it is incalculably destructive to the environment. Every single product that is manufactured is a subtraction from the natural world.

Gone are the days when things were built to last. There is a good chance that a desk built in 1900 is still around and has value. On the other hand, a desk bought in 2010, ordered online, made from particle board and assembled by the consumer is unlikely to still be in use. Most packaging, and even many products are designed to be one-time use. Cheap plastic junk is everywhere.

To fully address this issue will require significant changes in the current manner of living. At present the collective movement of society is in a diametrically opposite direction from that needed to reverse environmental degradation. Continued technological development is leading to an increasing need for energy as more and more people are brought into the Digital Age. The 11 percent of the world currently without electricity also want the goods of the digital age. Following will be a desire for phones and cars, and then better, smarter phones and cars. Forget almost every-thing else. The increasing demand for these two products alone presents an almost insurmountable obstacle to putting any checks on climate change or the elimination of environmental toxins. Now throw computers into the mix. And now all the other products that no one needs but must have. The problem seems intractable.

Corporate irresponsibility and the psychological inability of the human race to comprehend the significance of changes in climate due to short life spans, make it nearly impossible to shift behavior in such a way as to matter. The dependence on technology, and the belief that further tech innovation will reverse the tide, prevent even conceiving that the only solu-tion is radical change. Asking humans to give up their devices is akin to asking someone to denounce their religion. What percentage of people will voluntarily live a life comparable to that of pre–Industrial society in order to save the planet?

Unfortunately, the problem is so large that it requires coordinated gov-ernment action to address. The difficulties in expecting this to occur are readily apparent. Certain governments, such as China, are faced with the dilemma of pollution and the dilemma of economic development. Polit-ically, even in an authoritarian state, it is unlikely for governments to do what is necessary. The difficulty of getting democracies to take prudent actions can be seen in the withdrawal of the U.S. from the Paris Climate Accords.

Even if people can be made to go along and see the light, corporate reactions are predictable. Just as the oil and coal industries continue to push for their products and manipulate legislation in their favor, it can be expected that most, if not all, tech companies will do the same. We will continue to poison the environment, expend energy in ever increasing amounts and watch the world burn.

Even if we somehow manage to solve the problems of energy use and carbon emissions, the continuation of technological society will continue to produce toxic materials and their byproducts while causing further hab-itat destruction. A coming problem, one that has grown out of efforts to solve the burning of fossil fuels, will be that of battery disposal. At present, electric cars are not of large enough volume in the market to be an issue. However, as the technology develops and electric cars, or electric mowers, etc., become more common, the disposal of the toxic chemicals that com-prise this source of energy will become another issue.

In short, humans are killing everything else on the planet. As noted in the opening to this chapter the IPBES (2019) has estimated that one mil-lion plant and animal species are at risk for extinction. ONE MILLION! The report also indicates that at least 20 percent of the biodiversity of the planet has been lost in just the last century. In addition, half the total species on the planet are estimated to have died out in the recent past (Kolbert, 2016) and 60 percent of the biomass has died in the past 50 years (WWF, 2019). Arguments about whether we are in the sixth mass extinction seem settled. If half of the species have died and another million are at risk, clearly the planet is in a period of mass extinction. We are already there, and we are not yet through. We are not reversing the tide of destruction; we are not even slowing it down. At this point, it is rising ever faster.

Something must eventually give. Burning fossils fuels like there is no tomorrow will ensure that there is no tomorrow. At some point the human demand for energy and technology will exceed the capacity to produce them. What happens then? Has the planet already reached carrying capacity? As demand increases so will costs. Disparity between the haves and the have nots will also increase. Could climate and environmental change in conjunction with a greater demand for energy than the capacity to produce it lead to social disruption or a complete upheaval of society? Could technological society collapse? Will humans be able to adapt?

The question must be asked as to how rational beings can continue to behave as if everything is normal in the face of all the evidence. Are people so enamored of their devices that they do not care about the planet? Or is it that we have become so reliant upon and so impressed with our technology that we believe it can save us from disaster?

Technology has now become the mythical savior. No need to change our behavior, no need to conserve resources. Don't worry, be happy. Technology, our god, will save us. It is certainly easier to believe that than to take responsibility for our actions.

5

Unmet Promises

Life as a Beta Test

If technology is the answer to all problems then your pencil should be able to do math problems all by itself.—Rick Brantley

Popular culture often provides insight into the memes and expectations of a society. A popular futuristic cartoon of the 1960s, *The Jetsons*, reflects many of the unconscious ideas surrounding the development of future technology. The main character, George Jetson, goes to work in a flying car that collapses into a briefcase. He appears to work no more than a two-hour day with that work consisting primarily of pushing buttons. At home, food comes from a replicator and a robot maid does the housework. In short, the future was portrayed as one of fanciful machines that produced wealth without work, an abundance of free time and technology as the means to meet all needs. The technology of the future is almost always portrayed as miraculous, meeting all needs with little effort.

Interestingly, the problems of life never seemed to go away. George Jetson still had problems at work, his children continued to experience the problems of childhood and adolescence and the dog still had to be walked. A future of technological wonder on the other hand seemed accepted almost without question. It also appeared to be a given that all social ills had been solved.

The Jetsons, to be fair, was a cartoon aimed at children and intended to be happy and uplifting and as such should not be taken as an authoritative presentation of all media as regards the representation of technology. However, the meme of near miraculous technology is ubiquitous across science fiction, futuristic literature and media. The idea that the machines may not perform as desired is rarely considered. It is an article of faith that technology will work. The concern that there may be unintended consequences to technological advance is relegated to dystopian novels and subgenre movies.

Automation and machine mediated experiences are becoming normative, but the promise of increased leisure time seems less likely than ever before. Work does appear to be diminishing as a result of automation but the concomitant promise of wealth for all has not materialized. The upside of much innovation has been far less amazing than initially promised while the unpredicted downsides have at times wreaked havoc. No one envisioned at the time of the development of the automobile that in approximately 100 years the device would be a primary culprit in altering the entire ecosystem and climate of the planet.

Those old enough to remember the introduction of computers into common workspaces also likely remember the promise of the paperless office. The idea was that soon everything would be stored digitally and that there would be no need for paper. It was a great theory but not at all what happened. Instead, computers have allowed for the proliferation of printed material. Anything, including entire texts can be printed with little effort. As a result, draft upon draft along with countless things to read later get printed.

Much of this behavior is unconscious. Some individuals, many from an era when documentation was always on paper, routinely print everything they do from emails to work documents. Others print out articles they would never bother to read in hard copy form. Every person who makes copies makes a few extra "just in case." Others simply find it easier to read from paper than from a screen.

The paperless office is unfortunately not the end of the unrealistic and unfulfilled promises of both the digital and machine ages. A consequence of deifying technology is the belief that anything is possible and that every action must be carried out using the highest level of technology possible. Each technology is introduced with a promise of what it will do. Eventually the reality of what the new technology *can* and more importantly *cannot* do becomes apparent and is then accepted. Revisionist history will sometimes correct the promise to fit the reality.

The promise of certain technologies such as flying cars, cold fusion and transporters remain as illusive today as when first promoted. This optimism about what can be or might be prevents thoughtful analysis of what has already occurred. The ability to distort the past in a way that supports the views already held is a near universal cognitive bias.

The paperless office provides a convenient example not only of the unmet promises of technology but also the unconsidered and unintended consequences of technology. As noted, the paperless office has not really been achieved. Suppose for a moment that it could? Would that necessarily produce positive results?

The paperless office would, it is true, produce less paper. Would it use less electricity? Would the costs of digital storage long-term be more or less than paper storage of records? The answer to these questions is to some degree dependent on what is saved and how often it is accessed. For the NSA paper storage of documents is likely impossible (apparently so is digital due to the mountain of data being obtained) if not nonsensical. On the other hand, for a small plumbing company to upgrade or purchase computers, train personnel, etc., the costs might never be recouped.

In the area of commerce as specifically pertains to billing, it is easy to see how electronic payment saves resources for the company, bank, or other institutions. Not having to print out a bill/statement and mail it is an obvious savings for the company. For the customer, this appears to not be such a great bargain. If the customer wants a hard copy, for something such as taxes, the burden of printing and the attendant costs is shifted to them. The customer also must use electricity to view the bill even if not printing it and this is another cost passed to the customer. Unless one views their time as worth nothing, the time spent accessing documents online must also be considered a cost.

Technology often does little more than alter the form by which something is processed, not the actual task itself. Storing blueprints electronically does not build the house; keeping medical records digitally does not cure the patient. Apps that track physical activity do not, in and of themselves, burn even a single calorie nor run even one step. The collection of information is at present far advanced from the ability to utilize that same information.

Machines That Go Ping!

The belief in the ability of machines to outperform humans has led to a quest to infuse technology into all levels of society. In the area of medical care this practice has reached levels that have become counterproductive. The rapid transformation of the medical practice from a core group of patients seen by the individual physician to an interlocking system of providers and managed care networks has altered the means by which health care is delivered. This has fundamentally altered the relationship between the provider and the patient. Providers now spend more time entering data and managing files. Technology is increasingly being utilized to monitor the patients.

A visit to any hospital quickly demonstrates a peculiar effect this has had on the environment. The noise produced by various devices is almost constant. There is little to differentiate many of the tones produced by the machines. Some beep to show they are functioning properly while others only beep when there is a malfunction. So many alarms are happening simultaneously it is hard to determine what is a medical issue is and what is a technical issue. Some medical devices, such as the "machine that goes 'PING!'" in the Monty Python movie *The Meaning of Life* (1983), seem to make sounds only for effect.

However, the issue of sounds in the medical environment is not one of humor. The constant barrage of sounds leads to an unfortunate effect. For the visitor to a hospital the sounds are very noticeable and distracting. To hospital employees these become background noise. A condition known as "alarm fatigue" has been identified by the Food and Drug Administration (FDA) which documented over 500 deaths attributable to problems with hospital alarms in a five-year period (Tanner, 2013).

Over reliance on technology as a panacea for care has taken hold. So has an assumption that innovations will necessarily be good and without consequences. Many of the medical tests that were once considered absolutely necessary have now been shown to be unnecessary or even contraindicated. In just the past few years testing for breast cancer, prostate cancer and even routine gynecological exams have been called into question due to the problems of inappropriate diagnoses caused by overreliance on technology. It seems that in each of the cases once a technology existed to possibly detect something it became seen as completely accurate and the problem of false positives were ignored. Many people died as a result of being treated for something they did not have. Radical mastectomies, prostate surgeries, and many other procedures have been needlessly conducted by turning judgment over to machines. Interestingly, there is a backlash against data and research guiding medical care because numbers do not have the veneer of progress possessed by shiny new machines that have flashing lights and go ping.

An example of the refusal of the medical field to accept data and research—i.e., basic science—as opposed to reliance on medical devices—i.e., technology—can be seen in the difficulties found in gaining acceptance of a method shown to dramatically reduce surgical errors. The method involves instituting a simple checklist procedure to make sure that all necessary procedures have been completed before the patient leaves the operating room (Van Klei et al., 2012; Haynes et al., 2009; Lingard et al., 2008). Despite dramatic success rates in improving patient outcomes this approach is rejected by many physicians. It appears that its simplicity works against it (Van Klei et al., 2012). Many of the same physicians who reject following a proven medical approach because it does not seem sophisticated enough would likely support the spending of thousands, if not millions, of dollars to buy a new piece of medical equipment if it improved outcomes even marginally. Robotic surgery has followed the same pattern. Despite a number of emergent issues with these procedures, such as infections, this practice flourishes.

Creating Demand:
The Pusher Creates a New Addiction

Nautilus, that was the first gym membership I didn't use.
—Marc Maron

The comedian Marc Maron was likely on to something in the above joke. People continually make plans to exercise and get in shape. In the past, joining a gym was one of the ways in which people tried to make themselves exercise. The rationalization was/is that if you pay for something you will feel compelled to use it. Apps are the new gyms. There are numerous apps available to assist the nascent fitness buff in getting their fix. The price of these devices can range from free to hundreds of dollars.

Sports/exercise devices are but one of the many new products of the digital age designed to, or at least marketed as a means to, improve the health and life of the wearer. If only the devices could actually exercise for people. Wait—technology does attempt to do that. Tummy exercise belts are on the market that pulse and "exercise" the muscle with no effort—that being the gold standard in all technology—no effort.

It is unclear that a device that can monitor pulse rate can make anyone exercise. What is clear is that the search for magic solutions abounds in the world of technological innovation. The ability to develop technology can be seen to outstrip the ability to effectively and efficiently use that same technology. In some instances, the use of technology might be completely counter to the goals that are trying to be accomplished. An example of this might be the electric bicycle or the moving sidewalk. Efforts to apply technology for the sake of technology may have the purpose of undermining what is trying to be accomplished.

Is Technology to Blame for Our Failing Schools?

In almost all recent surveys of educational systems throughout the world, the U.S. is routinely found to be surpassed by numerous countries. How can this be? How is it that the country at the heart of the digital revolution, the homeland of Silicon Valley, the land of plenty, where everyone has two computers and a smart phone, can lag behind the rest of the world in education?

Ironically the answer could lie in the dedication to wiring schools and connecting to the world. Over the past decade there has been such a focus of putting a computer on every desk that everything else has been neglected. Teachers have been given mandates to introduce technology into the classroom yet accompanying support has been minimal. School budgets, often hiding behind bad economic news, have remained at best static at the very time computer demands have skyrocketed. As a result, teachers are furloughed, art and music programs are cut, and maintenance of infrastructure is deferred or ignored. Harsh economic realities over the past decade have essentially eliminated increases in education budgets and led to the elimination of many programs to meet necessary costs. Despite increases in population and increased social challenges that schools have been called upon to meet, funding has become static as can be seen in the amount of money spent on education in the Federal Budgets. It should be noted that for the same time frame listed in Table 5.1 the Department of Defense budget increased from $491 billion in 2004, when the country was involved in two separate wars, to a request of over $686 billion in 2019.

Table 5.1. U.S. Spending on Education, 2004–2019 (in billions)

Year	DOE Appropriation (a)	Total State Allocations (b)
2004	55.661	105.186
2005	56.577	109.111
2006	57.853	111.105
2007	57.483	119.199
2008	59.181	129.498
2009	63.524	158.152
2010	64.144	192.277
2011	68.345	184.508
2012	68.112	178.787
2013	65.704	172.941
2014	67.301	170.072
2015	67.135	164.298
2016	68.306	163.827
2017	68.239	164.631
2018	70.867	170.919
2019	71.448	171.537

a. Total Dept. of Education discretionary appropriation
b. Total Dept. of Education allocations to states

SOURCE: U.S. Department of Education, 2019. At www2.ed.gov/about/overview/budget/history/index.html

Numbers for the actual total dollars spent only on technology are difficult to calculate for several reasons. Matters such as the classification of technology in budgets, general funds, discretionary funds, etc., serve to obscure the actual amounts. Licensing fees and training for digital technology can be classified in numerous ways including administration, teacher training, and legal costs that further serve to hide the true numbers. However, what is abundantly clear is that at the same time overall budgets have remained flat or declined, the *only* category of funding that has increased over the past decade and a half has been for technology. Approximately a decade ago, the National Education Technology Plan (NETP), developed by the Department of Education, called for a "technological bombardment" as a means to "revolutionize" education (Richtel, 2011).

This investment in technology occurred during the "Great Recession," leading to decreases in every other area as total funding plunged. As a result, class sizes increased and money for other basic supplies suffered. Nicole Cates, co-president of the PTO at Kyrene de la Colina Elementary School, stated, "We have smart boards in every classroom, but not enough money to buy copy paper, pencils…" (Richtel, 2011). During this time frame funding for every category suffered. Teachers have gone decades with no appreciable raises as technology spending continues unabated.

The one factor that has consistently been found to most affect student performance and learning is class size. The one thing that would improve learning outcomes is the hiring of more teachers. Instead we've bought computers for every desk, are facing an infrastructure crisis, and have emergent security concerns that will prevent the hiring of teachers. The response has been to demand more accountability from teachers and demand that they be more efficient and effective. Does anyone really wonder why teachers report feeling less motivated and why so many leave the profession? All this and a political climate that skewers teachers as being overpaid and having too much time off might create a scenario where computers must educate children because there are no people willing to do so.

In addition to student and teacher needs, the infrastructure of the school system at large is also being ignored. According to the Center for Green Schools in the State of Our Schools Report (2013), it would require at least $542 billion to upgrade and modernize our schools to acceptable standards. Since the time of this report, matters have not improved. The need to upgrade the infrastructure of the schools just to reach minimal accepted standards does not bode well for the educational goals of the United States.

Perhaps the most disturbing aspect concerning the investment in technology of the educational system to the detriment to other factors in the process (yes, education is a process) is the lack of evidence to support its effectiveness. A review by the Department of Education (DOE, 2009) found that online courses "lack scientific evidence" of effectiveness (Richtel, 2011). Scientific research appearing in peer reviewed journals examining the issue of technology-enhanced learning has been less supportive with results described as "lackluster" (Sinclair, 2009), "exaggerated and unfulfilled" (Venezky, 2004), and even "ugly" (Dror, 2008). Larry Cuban, professor emeritus at Stanford, has been particularly critical of this investment pointing to "insufficient evidence," particularly as regards the money spent on technology (Richtel, 2011; Sinclair, 2009).

Despite the lack of scientific support for technological investment it appears there is no end in sight for this approach. The buzz in education is currently about "hybrid" and "flipped" classrooms which, it goes without saying, rely heavily on technology and online activities. Some social factors serve to encourage this approach even among those who know the evidence. First and foremost, the social meme, the current religion, which assumes technology to be an unquestionable positive, forces the argument. Being for technology is seen as being progressive about education and vice versa. What college administration will not be seduced by promises of more for less?

The ultimate expression of technology as the preferred means to deliver education (if one can deliver a process) is the emergence of MOOCs (Massive Open Online Courses). These extremely large courses are heralded as the wave of the future as they not only seem to solve the problem of too few teachers but also employ technology as the vehicle for content delivery. Seeing such online courses as an economically efficient way to deliver education content, schools such as Harvard and MIT have jumped on board. Economically efficient they may be, but they are not necessarily a guarantee of learning. In fact, looking at the evidence, MOOCs seem to exist primarily because they make money, not because they educate students. Often touted as free, MOOCs make money by charging fees for things such as having tests graded, reporting scores, etc. Tuition may, at times, be free, but it is a misnomer to identify the courses as "free."

In one course, on computational investing, more than 53,000 students enrolled but only 2,535 passed it. In another course with subject material in the area of electronics, 155,000 students enrolled but only 7,200 passed. In both cases, the majority of the students failed to make it past the first assignment or exam (Clark, 2013). Charging for services provided to only about 5 percent of those who pay for them is clearly a great way to increase profit, but not a very good way to educate students.

Online material is a great idea in theory. Unfortunately, what happens is that it becomes another barrier, another hurdle against delivery of material. The excuses are bad enough when students see the instructor every day; introducing a cyber element to the process adds another complication and another level of avoidance and excuse. Course management systems are mostly ignored by students. Full disclosure: this author is a college professor.

Beyond the issues above, the notion of MOOCs is doubly strange when one is aware of the literature surrounding the effective delivery of content in a teaching environment. Again, the one factor about the educational process consistently and repeatedly found to be the most important is class size. The smaller the class the better on almost every measure of outcome one can possibly imagine. The whole structure of the educational system demonstrates this implicitly and explicitly. At young ages children need small classes to get maximum attention and instruction. At the graduate level, students often work one on one with faculty because of the demanding nature of their studies. Being physically present in a learning environment is often a vehicle to discussions that are impossible in online classes. Interaction in hallways and student unions can provide pathways to knowledge that can never have the possibility of occurring in distance education or a MOOC. Education is not a commodity that can be purchased and consumed. It is a process that requires time and interaction. Technology often becomes a burden in that process rather than an a source of assistance. It is not impossible for a student to learn through MOOCs, but the whole process flies in the face of what is known about learning.

On a more anecdotal note, anyone who teaches at a large university is aware that a number of the professors are from other countries. In some fields, like math and science, foreign nationals predominate. What one often finds in discussion with them is that their backgrounds did not involve access to superior technologies and great schools but rather a dedicated teacher. An illustrative example is a math professor with an international reputation who attended school under a tree for the first six or seven years of school. He did not become a mathematician by having access to a mainframe but rather to a good teacher who gave him individual attention. His case is not particularly unique but demonstrates that access to the most advanced technology is not a requirement to learn. Teachers are the most important element in the process of education, not the methods and tools. Minds are not easily mass produced.

The Biggest Lie of All: Time Saving Devices

One of the most heralded advantages of technological development is a reduction of labor and an increase in leisure time. Machines are meant to free us of mundane tasks and take away the stress of work. If only the machines could update and fix themselves.

Since the invention of the wheel, humans have likely viewed every new technology as *the answer*. Every new gadget, be it a steam engine or a difference engine, is going to revolutionize the world. A brighter future is always just around the corner. Then the wheel breaks and we have to sit down in the middle of the road and try to fix it.

Have computers and other gadgets, or for that matter, any machine, really given us more leisure time? In this respect the washing machine may be the singular device that lives up to its hype.

Examination of an improvement to an existing technology will serve to illustrate how things have changed and whether this notion is true. Consider the typewriter. For many people, especially upon first introduction, the primary function a desktop computer serves is as a glorified typewriter. The word processing functions of the computer have truly altered the way books and other written media are produced. But is it any more efficient or less costly in terms of time?

In the past in order to write a paper, a person generally wrote a rough draft, made notes and edited the paper, sometimes making a second draft and then eventually typing a completed paper that might then be re-edited with only the changes getting a second typing. Some people, even then, composed at the typewriter, but still had to go back and edit and retype. Sounds like a pain, except when you think about the present. Mostly everyone composes at the keyboard. Editing becomes an ongoing process. Nothing ever seems done and the ability to print out draft after draft is easy. The final version becomes ever elusive.

When producing something required slightly more effort, things seemed more tangible. The advent of the copier was the first slip down the slope that equated ease of production to abundance of resources. This idea of abundance leads people to buy into the idea that technology is making life easier, more efficient, and better. What happened with the copier was not that it led to more efficiency in the workplace; it led to a proliferation of copies. Prior to the introduction of the Xerox 914 in 1959, approximately 20 million copies were made through the processes of duplication and mimeograph (Thompson, 2015). These machines required some operator knowledge and were cumbersome and messy. Only those of a certain age can remember the wonderful and likely toxic smell of purple ink on the paper. However, the ease of copying provided by the Xerox process transformed the way people managed documents and over 14 billion copies a year were produced by 1966 (Thompson, 2015). This represents a 700-fold increase in less than a decade. Desktop printers have not slowed this increase. Additive, or 3-D printing, will surely follow this course. The only factor that seems to limit printing is when the machines break down.

While printers seem to be designed to malfunction they are not the only device ever to break. Machines of all types have one common problem that is either ignored or glossed over in the initial stages of introduction. That problem is maintenance. All machines require it. This is true whether the machine is a wheel, automobile or computer. Everything breaks down and when this happens there are three choices: fix it yourself, have someone else fix it, or throw it away. In disposable society many things are not meant to be fixed and it is sometimes cheaper to buy a new one than to have it repaired. Even then, time is needed to find a replacement. Time. Isn't technology really supposed to free everyone from the mundane; give everyone nothing but leisure time? That is the promise. Is that how it is working out?

The automobile may move people from place to place faster than a horse and it certainly is nice not to have to shovel out a stable, but time is still required to buy gas, get the oil changed, have emissions tests, rotate the tires, check the antifreeze, wash it, etc. There may be a minimal savings of time in this comparison, but factor in costs such as pollution and suddenly the technological advance is not so great.

The automobile raises an interesting point in terms of our technological advance. We have gone from a state where new technology was usually built by the person using it to a place where the user has little, if any, understanding of the devices they use daily. Let us call the first stage of technological development the tool phase of technology. Simple machines easily constructed by the user. Examples of such tools would be the lever, the bow, or a wheel. In the second stage, let us call it the motor phase, technology required machines be built by someone other than the user. This stage generally required that machines be serviced by specialists, although with some understanding of the machine an individual might be able to service it themselves. Finally, we reach the modern era where our machines require a certain level of knowledge in order to make them work but when they break they are nearly impossible to fix. Let us call this the motherboard stage of technology.

In the motherboard stage our products rarely work when first acquired. Almost all the devices of the information age require attention to set up and keep running properly. In the early days of computers, one had to know how to write code in order to function. At a very minimum today one must know how to install updates and reboot the system. In the motherboard stage of technology everyone must become the equivalent of their own mechanic.

In a likely apocryphal story, Bill Gates reportedly once said that if the automobile industry were run like the computer industry cars would cost $100. Lee Iacocca, then the president of Chrysler is said to have responded that if the automobile industry were run like the computer industry cars would crash every three blocks (Brombacher, 2006). But what if the automobile industry were really run like the computer industry?

If cars were like computers the consumer would buy the car in one place, the motor in another, and the wheels in a third place. After purchasing the components, the end user then takes them home and assembles them in the garage. Then when the car does not start the person calls someone on the phone who then asked them if they have tried turning the key and putting gas in the engine. After three or so hours you reach a point where the person on the help line says "Hmm, I'm not sure why it's not working." After going through the whole process again, the customer is finally forced to call in an outside repairman. This repairman arrives and promptly insults the customer for not being able to install the motor themselves. After several hours of work, the repairman then needs to have the whole car and motor taken to his shop. Two days later he tells you that you have a faulty starter and need to take the whole thing back to the factory.

Sound far-fetched? Or does the above closely mirror problems that are common in the motherboard stage? Extend this a little further and think about how most people would function if the carburetor on the car were updated every six months and had to be replaced by the consumer.

One of the great hassles of the digital age, specifically as it relates to computers, computer programs and various other applications, is the constant process of updates. Certain programs seem to be in a state of constant update. It is the advancement by minutiae that is the characteristic feature of the digital age. Nothing is ever finished. Nothing ever seems to be released in a form that does not require debugging or constant tinkering. It is as if everyone is required to be a mechanic, a digital mechanic.

The race to market and the profit motive maintain this state of affairs. Companies are reluctant to spend time in product development out of fear they will be beat to market. The result is that the consumer is treated to a steady stream of minuscule, barely noticeable improvements that are marketed as "revolutionary." Meanwhile, technological advance is being used in a way that risks destroying the very things it is supposed to improve.

With each new technology there are trade-offs. In many ways technology just serves to shift the problem. Advancing technology has not prevented counterfeiting or theft, but it has shifted the means by which these crimes are perpetuated. The advent of email nearly killed the postal service, but it is still not possible to send a package as an attachment, at least not yet. With 3-D printers and drone delivery the need for the postal service will continue to be threatened. These trade-offs are important to recognize but rarely are they thoughtfully examined. To illustrate this point, examine the change and trade-offs caused by the introduction of automation into manufacturing. The level of production rose, but the displacement of workers caused serious upheavals in social structure. In the machine age this may have simply resulted in a shift of the labor force if one buys into economic arguments about "creative destruction," but in the digital age it is clearly reducing the workforce needed. How many other such technological trade-offs has humanity taken in its history that resulted in everyone becoming worse off, working harder, working more, competing with machines and becoming increasingly stressed and less satisfied.

We were promised the Jetsons, but we are becoming Blade Runner.

6

Unintended Consequences
The God of Technology Is a Tricky Devil

The spirit that I have seen may be a devil and the devil hath power t'assume a pleasing shape yea and perhaps out of my weakness and my melancholy, As he is very potent with such spirits, abuses me to damn me.
—Shakespeare, *Hamlet* (II, 2)

Scientific Illiteracy: Treating Science Fiction as Fact

Science fiction has loomed large in the background of technological development. From Asimov's *I, Robot* (1950) to Bradbury's *Martian Chronicles* (1950), to the whole of the *Star Trek* and *Star Wars* franchises, space travel has occupied the pop culture landscape. Humans in the far reaches of space are treated as a near certainty. Persons such as Elon Musk and Stephen Hawking talk of colonizing Mars or Titan. The tall tales of Baron von Munchausen (18th century) or the literary work of H.G. Wells (19th century) are the intellectual forebears of modern science fiction and increasingly science fact and technological development.

It is in the plots and stories of these science fiction works that technology is conceived. Much of what is considered the most marvelous technological breakthroughs have been inspired by devices and ideas that originated in the world of science fiction. One obvious example of this is the smart phone. Made wearable, as in the iWatch, it is only about one or two generations of development to Star Trek's combadge. The ability to teleport small pieces of matter has already become a reality (Pfaff et al., 2014).

The ability to think outside the metaphorical box of reality is one advantage of science fiction. Science is a different animal. In science, actual physical constraints do matter. Fiction frees one from such constraints.

Warp speed and time travel are convenient plot devices but (at least at present) unrealistic or impossible from a scientific perspective.

Consider the case of space travel. In *Star Trek* the mission is to explore and map the galaxy, "...to boldly go...." In *Star Wars*, the past was populated by an intergalactic web of nations and races. Perhaps there are places in the universe where species from different planets interact and travel from one nearby populated planet to another. For the inhabitants of planet Earth, the human species, that experience is likely to remain a fiction. The reasons are the vastness of space and the laws of physics.

The nearest star system to earth and hence the closest possibility of other life is Alpha Centauri at approximately 4.2 light years away. That means traveling at the speed of light it would take 4.2 years to get there. The speed of light is 186,232.397 miles/second (299,792,458 meters/second). The fastest humans have ever traveled is 39,897 km/h (24,791 mph) during the re-entry of Apollo 10. The fastest speed attained by an unmanned craft is the New Horizons rocket blast beginning the mission to Pluto with a speed of 58,536 km/h (36,360 mph) recorded during lift off (Guinness Book of World Records, 2014). Even if that speed could be sustained it would take around 78,000 YEARS for a craft from earth to reach the closest star (Byrd, 2017). Voyager I, launched in 1977, reached the Ort Cloud, the outer limit of our own solar system, only in 2012.

Even if great technological breakthroughs occur, the journey from here to the nearest system is easily 100 years or more. Given the lifespan of humans and the vast number of nearly insurmountable problems necessary to reach even the nearest solar system, where no evidence exists of any type of life whatsoever, encounters with extraterrestrial beings appears a nonexistent possibility.

Compound the issue of the relatively slow pace of travel thus far attained by humans with the problems of heat and friction associated with increasing speed. As speed increases so does the heat associated with travel. In other words, getting to a speed anywhere close to what is needed produces additional hurdles that need to be solved in order to not burn up the vessel. The result is that space travel beyond our own solar system appears unrealistic in anything like a foreseeable future. Add the burden of providing a living environment for humans and the questions of what might be encountered during space travel, and ideas such as journeying to another star system seem a fool's errand.

A multitude of other issues also render the idea of extended space travel as more in the realm of fiction than fact. There is a noted physical effect upon the eyes that causes a decrease in acuity at around six months in space. Studies to examine the effects of longer space exposure on the visual system are necessary before extended flight. Human eyes did not evolve in zero gravity and it may be that living in such an environment will produce irreversible damage. The issue of the effects of long-term exposure to space radiation upon the human brain may also prove an insurmountable challenge (Khan, 2016). Add in the problems of food and water storage as well as the difficulties of living in extremely confined spaces for extended periods of time and it may well be that humans are not physically up to the job. The relatively simple task of putting a satellite into orbit is still not a sure-fire proposition, as the explosion of the multiple SpaceX rockets attests. One satellite was intended to bring Internet service to extensive regions heretofore unreached while another was to supply the International Space Station (Chang & Isaac, 2016).

The fascination with manned space travel is one potent example of the means through which humans self-aggrandize through technology. Human presence during space exploration is just a burden in terms of resources and design. The accompaniment of humans along for the ride tends to hinder what can be scientifically studied and limit the mission as regards duration and flexibility. Rather than room for scientific equipment the need to create a suitable environment for living becomes paramount. Manned flight is ultimately resource wasting. Any effort to reach beyond our own solar system must acknowledge this reality. If there is ever a time and place for robots, space exploration is that time and place.

But let us take a leap and consider that all the problems with space travel can be solved. Even if the answer to space travel is that it can be done, the very real question remains as to whether it should be done. Ethical issues as regards resource allocation, psychological attitudes towards conservation, human adaptability, and multiple other concerns must be addressed in addition to the purely technical questions. These matters are not typically within the realm of concern for most inventors or technophiles.

It is not an overstatement to say that most inventors and IT professionals have little understanding of philosophy. In a nationwide survey of degree requirements, it was noted that those in information technology and related fields such as engineering tend to have relatively higher degree requirements in terms of total hours when compared to other fields. However, attention to ethics as reflected in these requirements for courses in the subject is virtually nonexistent. Most schools have at best one departmental course in the subject while many have none at all (Johnson, Reidy, Droll & LeMon, 2012; Pitter, 1996). Many course "requirements" in the area of ethics actually come in the form of electives taken in the general studies program (Gorgone, Davis, Valacich, Topi, Feinstein & Longenecker, 2003). Other assessments of ethics courses in the area of information technology found these courses to be nearly nonexistent (Xia, 2018).

This very lack of scientific and liberal arts training among the inventor and technologists' classes has had and will continue to have profound impacts upon all of humanity. The severe environmental degradation directly and indirectly attributable to technological advancement is but one example of the problems unintentionally caused by innovations that were to improve the quality of life.

In the early 1900s the most sought-after design feature in watches was the ability to make the dial light up so that it could be read at night. The process was simple and involved the application of radium directly onto the face of the watch. Several decades later it was determined that the workers who made the watches died of cancer and other diseases directly related to their employment. The process of painting the watches apparently involved licking the brushes used to apply the radium in order to keep them properly trimmed and as a result the workers directly ingested the hazardous material (Mullner, 1999).

The temptation is to dismiss this as an aberration, a one-off stroke of bad luck. That would be an incorrect surmise. Many of the scientists involved in the Manhattan Project were reportedly haunted by their role in developing the bomb. It was only after the fact that they began to see the enormity of what they had done. It is impossible to foresee the future and the unintended consequences of any invention can never be fully known.

Previously it was noted that those in technological fields often have little exposure to philosophy and ethics. Coursework in these fields is largely viewed as an impediment to study, something to endure when it is a basic requirement rather than something that is essential. As poor as the training in ethics and philosophy is for those in technological fields it may be even worse in the world of business. Combine the two and what emerges is a drive to develop products first and worry about the problems later. Lack of adherence to guidelines concerning precaution in the business community serves to make new technology untrustworthy from a health and safety point of view. What you don't know can kill you. The history of human invention, and by extension technology, is filled with individuals who died as a direct result of their own inventions.

Table 6.1. Individuals Killed by Their Own Discoveries/Inventions

Perillos of Athens (c. 55 BCE)—developed a method of execution known as the "Brazen Bull." He died when a demonstration of the device was demanded.

William Bullock (1813–1867)—inventor of the web rotary printing press, he died from a resulting infection after his leg was crushed in the machine.

Julius Kroehl (d. 1867)—developer of the sub-marine Explorer, the prototype of a submarine; he died of decompression sickness.

Otto Lilienthal (1848–1896)—inventor of the glider.

Sylvester Roper (1823–1896)—developed the first motorcycle; the Roper Steam Velocipede.

Aurel Vlaicu (1882–1913)—builder of the first metal airplane, he crashed in the Carpathian Mountains.

Valerian Abakovsky (1895–1921)—developer of the high-speed rail car, he died during derailment that killed him along with six passengers.

Marie Curie (1867–1934) and Pierre Curie (1859–1906)—Radium got them and everybody else in their lab.

Sabin Arnold Von Sochocky (1883–1928)—developer of radium luminescent paint, he died from aplastic anemia as a result of radium poisoning.

Max Valier (1895–1930)—built a rocket powered car; during a test, the motor exploded.

Manhattan Project (circa 1940s)—Scientists working on experiments—Harry Daghlian (died 1945), Louis Slotin (1946).

Henry Smolinski and Hal Burke—built the AVE Mizar, a flying car that combined a Ford Pinto and a Cessna; during a routine flight the wings fell off and killed both.

James Heseldon (1948–2010)—Segway: he didn't invent the device, he just owned the company. Killed when he fell off one of the devices.

Opening Pandora's Box

Steven Johnson, in *How We Got to Now* (2014), delineates several examples of the rather unexpected and capricious way inventions have impacted society. Specifically, he focuses on the way in which an invention can impact areas never before considered. The printing press, which placed written material in the hands of the masses, provides a clear example of this concept. The wide dissemination of reading material also led to the discovery of visual problems and the invention of glasses. In turn this led to microscopes, telescopes and the scientific revolution (Johnson, 2014).

The expression "opening Pandora's box" also has roots in the philosophy of the unintended consequence. Yet, in the world of technological innovation, it appears this concept has never been considered. Every problem, as noted, is perceived to have a technological solution. This is true even when the problem was originally a solution.

Pollution and the degradation of the natural world are palpable examples of some of the unintended consequences of technological development. Efforts to solve these problems through other technologies such as nuclear power and hydraulic fracturing have exacerbated matters.

Unproven technologies and their unknown effects are not the sole cause for concern. In a bit of collective denial, computer users and other consumers of high-tech products routinely ignore the very manufacturing process that leads to their shiny new objects. Sweat shops, slave labor, child labor, working conditions that expose workers to toxins, and more, are a ubiquitous part of the digital age. It can correctly be argued that exploitation is not a by-product of technology. However, the ethical implications of poisoned workers are easier to ignore when it happens out of view. What is abundantly clear is that technology has now provided the potential and possibility of causing havoc on a much larger scale.

A conundrum of technology is that many of the things that can and do make real improvements in the lives of people, such as prosthetics for amputees or children with birth defects, inevitably get extended to military applications. Simple prosthetics become a step to robotics, then to bionics, before ultimately becoming robot soldiers and then killbots. The unfortunate natural progression in technological development seems to follow the course of: Initial promise → Application with good intent & purpose → Repurpose & Improve → Military Application

Occasionally something follows the opposite course with military development first and then the civilian applications. The computer and the mobile phone might be examples of this reverse course of development but even at their core violence must be associated with these technologies. Mobile phone technology originally came into being as a means to guide torpedoes. In an interesting quirk of history this technology, called spread spectrum, was developed by actress Hedy Lamarr and is also used in Wi-Fi (Women Inventors Website, n.d.).

Even when the original purposes are military, the violence associated with the technology is brushed aside as essential for the greater good. Beginning with objectives that are nearly impossible to question, most technology is seen only through the lens of what it can do to better the human condition. The downside, or the unintended consequences and effects are rarely considered. In many ways it is almost impossible to determine where technology will eventually go. For much of technology, the military applications come later, often developed in secrecy. This is problematic precisely because it is impossible to know the future. A seemingly unambiguous good might turn out to be instrument of malicious behavior.

The Dark Web

When the Internet was being developed the primary concerns were how to develop the protocols and infrastructure necessary for the system to function. At the time, communication and exchange of information were the primary objectives. Commerce came later. What appears not to have been contemplated nor intended was the spate of illegal activities that are facilitated by connectivity.

The first type of illegal activity that drew significant attention involved file-sharing, primarily of audio files, on sites such as the now defunct Napster. Incidentally, Napster died when it was bought by the massive conglomerate Bertelsmann Music Group (BMG). File-sharing has nonetheless continued and the downloading of files for music, movies, and other entertainment is firmly entrenched in the habits of netizens. Software such as BitTorrent and other similar methods make this so easy that many users likely have little understanding of the illegal nature of much of what is done. Shutting down any specific site does little to prevent or even slow behavior as copycat and mirror sites, those with the same information but at a different IP address, provide the same content.

In truth, file-sharing is the least of concerns when it comes to illegal activity. From a certain point of view file-sharing might not be ethically wrong even though illegal. Why is loaning a digital copy of something different from loaning a physical copy? Some artists see it as free advertising and make their own content available without charge. On the other hand, some companies prosecute every 11 year old who downloads a pop song.

The most infamous of sites dealing with illegal activity is the now shuttered Silk Road site. The site was an open market for drugs, prostitution, forged passports, hacking software, and many other similar activities. The site itself was buried in layers of encryption and web pages that made it difficult for law enforcement to determine the operators. Difficult, but not impossible. The operator of the site, Ross Ulbricht, who used the online sobriquet "Dread Pirate Robert" (from the *Princess Bride* movie) was eventually arrested in the public library in San Francisco (Flitter, 2013; Goldstein, 2013; Grossman & Newton-Small, 2013). Although Ulbricht received life in prison with convictions ranging from solicitation of murder to complicity in overdose deaths (Neumeister & Pearson, 2015), his arrest barely slowed the flow of illegal activity on the web. A proliferation of other sites quickly arose to fill the void.

While Silk Road specifically prohibited child pornography and counterfeiting activities, other sites do not. A site called WHMX specializes in counterfeit currency (you probably have to pay in Bitcoin) while other sites specialize in particular crimes like passport forgery or prostitution (Grossman & Newton-Small, 2013).

Child pornography has proliferated in the digital era and is aided by Internet anonymity. Common apps can be used to track individuals for cyber stalking. Laws governing illegal activity in cyberspace are woefully inadequate in that most crimes must be prosecuted in the jurisdictions where they occur. Child pornographers and others with malevolent intent often set up shop in places where regulation is nonexistent in order to remain out of the reach of law enforcement. To catch such individuals may require international efforts. Interpol has become involved in the battle against child pornography successfully taking down several sites including one called "Freedom Hosting," purported to be the largest supplier in the world. The owner of the site, Eric Marques, had to be extradited from Ireland (Grossman & Newton-Small, 2013). The hacking collective Anonymous has also taken down several child pornography sites through its activities (Kushner, 2014). The enormity of this problem can be seen in an investigative report that found 45 million photos and videos categorizable as child porn in the previous year (Valentino-DeVries & Dance, 2019).

The degree to which content is hidden or buried on the web is surprising. A survey of the web in 2013 found that of the approximately 7,500 terabytes of information that were believed to exist, only about 20 terabytes were indexed (Grossman & Newton-Small, 2013). While the absolute numbers of information on the web constantly changes the fact that only a small percentage is indexed is obvious and indisputable. In addition to outright criminal activity, social media sites for terrorist groups, organizations associated with criminal activity such as street gangs like the Crips and Disciples and motorcycle gangs like the Hell's Angels, and even fronts for hit men and assassins for hire, are all available. Worrying about file-sharing seems tame by comparisons. Researchers studying the cyber world and its darkest corners have been found to experience symptoms of depression and helplessness after dealing with these issues where a constant stream of toxic information is the norm (Marwick & Lewis, 2017).

Outright theft through such means as hacking, identity theft and fraud have previously been discussed and not repeated here. It is obvious, however, that the ability of the digital world to become an accomplice in criminal activity was not an intention at the conception of the Internet. The cyber world has created a totally unexpected boost to criminal activity. Current laws and jurisdictional constraints leave many crimes outside the boundaries of prosecution. The resources required to unravel the anonymity and then infiltrate criminal sites makes all but the largest and most visible offenders unlikely to face much of a legal threat. By the time Silk Road was busted it was well known and the subject of media interest. It is an irony of the digital age that anonymity can be a business model.

Misguided Men: Bad Actors Redux

Our scientific power has outrun our spiritual power. We have guided missiles and misguided men.
—Martin Luther King

The attitude that technology makes everything better is rooted in the adoption of technology as a religious way of life. Believing that technology can solve all problems leads to a viewpoint that disallows any questioning of the approach. Suggesting that technology could be used for bad purposes is viewed as heretical. The experience of the arms race would appear to be prima facie evidence of the dangers of unthinking technological innovation. However, as with all religions, contradictory evidence is excluded and refuted.

Developers of technology never seem to consider the misapplication of their inventions. Advances in 3-D printing appeared nothing short of miraculous right up until someone printed a gun. Even something as innocuous as a laser pointer has the potential to be misused. In a report the FAA noted that the use of laser pointers aimed at airplanes was up 1,200 percent (FAA, 2014; FBI, 2014; Kerlik, 2014). It is something as ridiculous as these incidents that raise a real alarm about technology that appears overlooked by inventors—namely that there are those with nefarious purposes. Cyber thefts, cyber espionage, cyber stalking are but a few of the problems created by the online world. The use of technology by terrorists and criminals to evade detection is another example of the problems created by good intentions.

The ability to build robot soldiers is promoted as a great boost in the waging of war. In effect, the idea is promoted that no one gets killed. No one on "our" side, that is. Drones are just another extension of this attitude. The neutron bomb, killing people but not destroying buildings, might be the ultimate expression of this attitude.

All of these serve to sanitize war. Does this not make war seem less horrific? Might this not serve to make war easier to sell to the public? The Iraq War had some of this. Through superior firepower and technology (the ability to see at night for example) the U.S. would, according to Dick Cheney, "just sweep the Iraqi's aside." It did not work out that way but the idea of a war without soldiers being killed began to take hold. Attempting to contain the outcry over the war, the occupation forces increasingly relied on machines as a means to keep U.S. casualties low. Drones became the preferred method of attack and turned the war into a video game for the pilots. The era of robotic war was slowly ushered in but now is entrenched. The U.S., China, Russia, Israel and Germany are already moving towards the development of robot soldiers according to Human Rights Watch (2016). Britain is one of the few nations that categorically denies that it is working on anything in this area. The technology to develop and utilize autonomous robots may already be possible and if not it is assured that this eventuality is not far away. The ethical issues raised by the specter of robotic and cybernetic soldiers are enormous. If both sides have robot soldiers (killbots), it just seems to be a pointless waste of resources. If only one side has these weapons then it is slaughter. Asymmetrical slaughter due to superior firepower does not exactly fit the definition of a "just war."

Moving beyond the question of automated war, there remains the issue of the irrational actor. Those who use technology to steal, wage war, engage in spying, etc., are all cases where the motivations can be understood and possibly countered. However, even if the machines were made fail-safe, the person who behaves irrationally often cannot be countered. Take the case of 9/11 or the cases of all the mass shootings that end in suicide by self or cop. Guns and bombs are all inherently violent technology of whose use and control is predicated on the notion that all people are responsible. This notion is patently false.

The possibilities that remote controlled or autonomous killbots create are limitless. On normally staid public broadcasting news programming, serious analysts openly contemplated the possibility of nanoscale killbots and assassination drones disguised as insects that could inject poison with a micro-needle and then self-destruct leaving no trace (McLaughlin Group, 2015, April 26). Surely, multiple terrorist groups are seeking to obtain autonomous vehicles as weapons delivery systems.

Other possibilities lie in technology heretofore unknown. The case of U.S. diplomats in various embassies bears some examination. Serious health effects among a number of embassy staff have been detected. Although first dismissed, the health issues now are documented and there appears to be some consensus that the problems have some connection to a sonic or sound-emitting weapon (Dyer, 2018; Skopec, 2018; Tumolo, 2019). Much of the information surrounding these events appears shrouded in secrecy. However, it is clearly suspected that some form of state espionage or spy craft is at work.

Couple this with the knowledge that there are world leaders with access to nuclear weapons who seem to be in questionable mental states and the world is seen as undoubtedly less safe as a result of advancing technology. One starts to think about the existence of people who have a mindset that might appropriately be called psychopathic, who seek to destroy the world, or at least some corner of it, just for the sake of doing so. Unfortunately, even if all the bad actors could somehow be caught or controlled, it is still possible that if everything works as planned all might not go well.

All Devices Need Energy

One indisputable fact of the digital age is that it runs on electricity and fossil fuels. Even with efforts to improve efficiency through numerous means energy use worldwide continues to rise year after year. This has occurred even as the largest consumer, the United States, has shown relatively flat usage patterns over the past decade with total consumption peaking in 2007 and holding steady since then (U.S. Energy Information Administration: EIA, 2014). In the rest of the world, however, this is not the case.

Total world consumption of energy rose steadily from 491.9 quadrillion BTU in 2007 to 558.7 quadrillion BTU in 2014 (EIA, 2014). Estimates provided by World Energy Projection System Plus indicate this amount will steadily increase reaching 819.6 quadrillion BTU by 2040 (International Energy Outlook, 2013). None of this should really be surprising as the rest of the world seeks to acquire the goods and services of the technologically developed world. Automobiles are proliferating in China and India while cell phones are quickly connecting the whole world. Efforts to wire the world and provide a computer to everyone on earth only compound the problem of energy consumption.

This problem, of excessive energy consumption, is obvious. However, the unconsidered and unintended effects of this excessive consumption are only now becoming apparent. According to calculations by eminent biologist E.O. Wilson, it will require four more earths in terms of natural resources for everyone on the planet to live like those in the U.S. (Wilson, 2002). Yet this standard of living is precisely the goal in the rest of the world.

Something has to give. The solution to the problem of energy consumption has thus far been in the realm of magical thinking. It is easier to buy into the dogma of the Church of Technology that a solution is just around the corner than to make any efforts to change human behavior. Such beliefs are nothing more than pie-in-the-sky promises. All such efforts tend to ignore the problems any new solutions might produce while overestimating their efficacy. Nuclear energy and fracking spring to mind as ready examples. The lurking battery disposal crisis will become yet another solution turned into a problem. Yet belief in technology to fix all problems continues to send the planet spinning towards unsustainable actions and behaviors. Technology will save us, is the mindset of the techno evangelists and their followers.

Particularly frustrating is that there appears to be little the individual can do to alter this continual energy crisis. According to data from the EIA, most use is beyond the control of the individual consumer. Industrial use of energy is estimated to be 51 percent of total consumption while commercial use accounts for another 12 percent. Transportation uses around 20 percent and residential use is surprisingly low at 18 percent (EIA, 2014). In other words, without broad policy and social changes there is little evidence to expect anything but continued increases in the demand for energy. Some of this wasting of energy is unintentional but at another level it is an expression of faith in technology to solve problems. The wasting of electricity has become institutionalized in the digital age.

In a report authored by Mike Mills of the Digital Power Group titled *The Cloud Begins with Coal: Big Data, Big Networks, Big Infrastructure, and Big Power* (2013), a wealth of information is presented that outlines the true costs of the virtual world. When energy needs are considered, even virtual items that would seem to have no carbon footprint turn out to be remarkably damaging to the planet.

For instance, Mills found that it takes more energy to stream a movie from the cloud to a consumer than to manufacture *and* ship that same movie to the consumer's house. Another example, that seems unbelievable until one views the actual numbers, is that it requires more energy to power a smart phone (361 kWh/year) than a refrigerator (322 kWh/year) (Mills, 2013). This calculation takes into account wireless connections, data transfer and usage, and battery charging drains of electricity. Add to this the fact that while few homes have more than one refrigerator, many, if not most, have more than one smart phone. Phones are not the only culprit. A walk through almost any home at night confirms a well-known feature of the digital age—nothing is ever completely off.

Many digital products, even when off or asleep, have a standby mode. For many products, these modes use almost as much energy when in standby as when being operated. The Lawrence Berkeley National Laboratory has calculated the various rates of uses for various devices.

Table 6.2. Rates of Electricity Usage for Various Devices While in Standby Mode

Device/Product On	Ave. Watts	Maximum Watts	Ave. Use While in Standby Mode
Mobile Phone—charge	2.24	4.11	3.68
Computer display CRT—off	0.8	2.99	65.1
Computer display— sleep	12.14	74.5	65.1
Computer desktop—off	2.84	9.21	73.9
Computer desktop—sleep	21.13	83.3	73.9
FAX—inkjet	5.31	8.72	6.22
Modem—DSL	1.37	2.02	5.37
Modem—cable	3.84	6.62	6.25
Phone—cordless	0.98	1.8	1.9
Printer—inkjet	1.26	4.0	4.93
Set top box—DVR	36.68	48.6	29.29 (Record)

Device/Product On	Ave. Watts	Maximum Watts	Ave. Use While in Standby Mode
Set top box—cable	17.5	26.3	29.64 (TV On)
Set top box—satellite	15.47	32.7	16.15 (TV On)
Stereo—portable	1.66	5.44	3.3
TV—CRT	2.88	16.1	N/A
CD player	5.04	18.4	9.91
DVD/VCR	5.04	12.7	15.53
Game console—off	1.01	2.13	26.98
Game console—ready	23.34	63.74	26.98
Surge protector	1.05	6.3	0.8 (on)
VCR	4.68	9.9	7.77

SOURCE: Lawrence Berkeley National Laboratory. Adapted from Standby Summary Table, at standby.lbl.gov/summary-table.html.

The table demonstrates some very bizarre facts. For almost every product, there was a version that used more energy in off/standby mode than was the average for the same product category while in operation.

These facts present a very stark reality. The explosion in energy use is not caused by people in developing nations getting automobiles (although that fact is a nightmare as regards carbon emissions and climate change), but rather by digital devices in the developed and developing world.

A particularly troubling symptom of this belief system is the conversion of formerly manually operated objects into those requiring electricity. Electric can openers and shavers were some of the first harbingers of this trend. Electric bicycles (seems to defeat the point, doesn't it) are one of the more recent examples of this approach. The nadir of this trend has to be the conversion of objects to electric operation that seems nonsensical. The electric trashcans at the Atlanta Hartsfield Airport—the world's heaviest trafficked—are a prime example. Moving sidewalks, present at almost every airport and many public buildings, are just needless wastes of energy. It is a particular form of delusion that views wasting resources as progress simply because an object is now high-tech.

The problems of climate change and pollution are likewise being met with efforts to fix the problem by technological innovation. Efforts, that is, to find a technological solution. Attempting to address pollution and excessive energy use are apparently unacceptable as public policy. Instead, efforts to alter entire ecosystems are presented as reasonable. These efforts are collectively known as geo-engineering.

Geo-engineering

Geo-engineering can be defined as the deliberate effort to alter and manipulate the environment on a massive scale in order to address the effects of climate change. It involves such actions as cloud seeding to alter the weather, creating mechanical "sunshields," using chemicals to block the sun, attempting to filter the air and numerous other methods to alter the environment (Calderone, 2015). Ideas involving efforts to alter the climate indirectly by methods such as dumping iron dust into the ocean to encourage the growth of organisms to absorb CO_2, seem motivated by a need to "do something" rather than address the cause of the problem.

The fact that these ideas have even been conceived at least points to awareness of the critical state of the environment. The fact that the IPCC brought up the idea in an annual report (IPCC, 2014) also points to a type of desperation on the part of those aware of the issue of climate change. However, desperation rarely leads to good decision making. Any reasonable cost-benefit analysis would quickly kill any such idea because the potential risks should something go wrong are too great. Geo-engineering falls squarely in the realm of faith-based science and reflects an admiration for high tech solutions for all problems—especially those caused by technology. The most extreme example of this type of approach involves using a nuclear weapon to alter the orbit of the earth in such a way as to make it further from the sun and therefore cooler (Falconer, 2014).

Efforts at geo-engineering appear to require massive and coordinated efforts even to attempt. Although that would seem to be the case, it is not necessarily true of all the methods. Despite international agreements prohibiting such actions, Russ George, a California businessman, dumped 100 tons of iron filings dust into the ocean to see what would happen (Doyle, 2013; Fountain, 2012). The fact that one individual could pull this off is, to say the least, alarming. The long-term effects of this action have yet to be determined (Johnson, 2013). Even attempting to determine all the relevant factors involved when altering an ecosystem is impossible.

Harvard, in an effort led by David Keith, is also studying the feasibility of geo-engineering using some form of cloud seeding (O'Brien, 2019). Acknowledging that the effort could make things worse, the project is still proceeding. In many ways the true danger of this effort is that an effort with the backing of a respected university might be allowed more leeway than would otherwise be warranted.

The major issue with geo-engineering is the possible enormity of the consequences. Blocking the rays of the sun might seem like a good idea until it goes too far or becomes irreversible. This approach, for example, might produce something called the "termination effect" which could lead to spikes in the levels of heat and make things even worse. The whole idea that a risky and poorly understood approach should even be attempted speaks more of desperation and irrational hope than reason. Even actions that are effective may produce undesired reactions in other parts of the planetary ecosystem. Rather than address the underlying causes of climate change, rapacious energy consumption and its concomitant pollution, the mindset is to change nothing about human behavior and count on future technology to reverse it. In many ways the greatest danger of efforts to engineer the climate was pointed out by Miles O'Brien of PBS News, who stated, "If we think we can do it, it will make us rely on it."

The real climate engineer is human activity. That is what must be addressed to reverse the course of environmental degradation. The Union of Concerned Scientist (UCS) has produced a low-tech plan to save one-half of all energy consumed and may be viewed on their website (UCS at www.ucsusa.org/). The plan involves little in the way of technological change and focuses more on the behavior of people. Attention to the plan has been nearly nonexistent.

7

Involuntary Cyborg

No Place to Hide

Technology has become as ubiquitous as the air we breathe, so we are no longer conscious of its presence.—Godfrey Reggio

In the science fiction series *Star Trek: The Next Generation*, one of the most powerful foes encountered by the crew of the *Enterprise* was a race known as the Borg. The Borg were a technologically advanced society that continually modified and advanced itself through the assimilation of other cultures and their technology. Governed by a queen and controlled through the collective hive mind, the Borg did not actually invent any of its superior technology; it took it from the other civilizations it encountered, conquered, and assimilated. Veni, vidi, vici, vivi(sectioned).

Operating with machine-like efficiency the Borg plundered the universe, imposing its will on all with whom it made contact. Resistance, the Borg warned, was futile. Crushing the other cultures it encountered, the Borg collective never stopped to consider the implications of its actions. It saw itself as superior and was certain that its actions improved every culture that it assimilated, i.e., enslaved. The logic of its actions were self-evident to the members of the collective.

So it is with the digital age. Resistance is futile. Like those assimilated by the Borg, there may be no possibility of an alternative. In the technologically advanced, so called developed world, it is nearly impossible to exist without at least minimal connectivity. Even people who view themselves as living an "organic" existence and who brag about their low carbon footprint can scarcely exist without access to a computer or mobile devices.

To participate in society, one must accept and acquiesce to technological advancement. Assumptions and priorities promoted by the deification of technology lead to a bias towards newer and increasingly complex mechanisms. As a result, upgrades are a fact of life. Implementation of these "advances" is without debate and what is defined as progress occurs beyond conscious consideration. Even for those who have some ability to distance themselves from consumer culture there is little recourse. In order to keep the various devices, programs and apps running properly, upgrades are a necessity.

It is in the realm beyond simple upgrades that technology most directly impacts life. Technology is currently reshaping nearly every aspect of life in a manner analogous to the beginnings of the Industrial Revolution. Remote and mobile controlled devices are beginning to run homes and automation is becoming an expectation. As technology progresses and begins to augment, complement, enhance, and otherwise alter humans in the manner in which machines have been altered in the past, the very nature of what it means to be human will be challenged. Conscious debate concerning the direction and limits of the merger between humans and machines is needed.

For most of the products of the digital age choice is largely illusory. Debates over operating systems, mobile services, etc., are insignificant when compared to the debate about the direction of technological development. This debate in a practical sense appears largely decided by a digital high command, Morozov's (2013) "digital elite," that drives an increasingly technological future. This future is promoted as *the only* way forward. Those who question the direction of development especially as regards matters such as merger with machines and artificial intelligence have about as much chance of being heard as a Star Fleet Officer trying to reason with the Borg. The consequences of accepting a unilateral drive forward could be similar to an encounter with the Borg—the true power of the collective is not known until all has been subsumed. Might humans be rendered obsolete?

Killing the Goose
That Laid the Golden Egg

Creative destruction is the economic term for the loss of jobs caused by technology. As has been shown, the effects of automation on the labor market have had a severe impact on human employment. In the past, creative destruction led to humans shifting from one type of job to another—e.g., stable hand to mechanic. In the present, the human is replaced by a machine. Unfortunately, this direct replacement of workers is not solely limited to manufacturing jobs.

In the realm of what might loosely be called artistic endeavors, technology is beginning to have great effects that seem wholly unintended but threaten the existence of entire industries. In the area of recorded music, most media attention has focused on file-sharing and its economic implications. Music company executives have decried this as violations of copyright. However, many musicians have embraced file-sharing as free advertising. In the new music economy, some artists, such as Curtis Jackson, aka 50 Cent and Chance the Rapper, seized control of the means of production and distribution, cutting out the middlemen. Chance's Grammy Award winning song in 2017, for example, was available only online. The issue of copyrights aside, digital music has other challenges.

Multiple apps allow the user to identify songs and also allows for a Big Data approach to music. Utilized as a predictive program by companies to determine what will be hits, the result has been to undermine creativity and bury the work of lesser known and new artists. Online revenue from streamed music demonstrates this effect with the top 1 percent of artists generating 77 percent of revenue (Thompson, 2014). This has the further effect of leading music companies to look for more of the same. As a result, music has become more repetitious and blander. This is not idle conjecture. A study conducted by the Spanish National Research Council (SNRC) bore this out finding that music is demonstrating "less variety" but increased "loudness" (Serra et al., 2012, Thompson, 2014).

Books have followed a somewhat similar course in the crushing of the artist and small businesses. Amazon.com and online sales have come to dominate the sale of books to such a degree that brick and mortar establishments have become endangered. This affects not only the businesses but also the consumers and by extension the writers. Consumers no longer have the ability to peruse books and become acquainted with sales personnel who might be aware of titles the consumer would not find online.

For writers, the ability to have any assurance of copyright control is essentially gone. Who knows if digital accounting is accurate? It likely does not matter as compensation in the form of royalties from digital books is substantially less than from physical copies. As with musicians, writers see little in the way of revenue from digital downloads. For the publishers and online sellers, e-books are a marvelous product in that there are almost no production costs, low royalty rates, and high profit margins. It matters little that the consumer prefers tangible books, with evidence suggesting that people read e-books at a lower rate and eventually develop full e-bookshelves at which point they quit buying e-books as well.

Beyond the issue of digital products, physical products online have concerns with regard to quality and integrity. Likely anyone who has ordered more than three products online has experienced some degree of frustration over the product that arrived. It is very difficult to determine quality in cyberspace. A good photographer can make polyester look like silk.

The world of work is one of the more evident examples of technology proceeding like a juggernaut across an entire sphere of life with little regard for its ramifications. Many labor and service functions formerly conducted by human beings have been automated. The replacement of manufacturing jobs by automation initially happened slowly and out of sight and as such escaped conscious recognition until it began to cause significant disruption in the labor market. This period, from the 1800s through to the early 1900s was marked by alarm over the effects of machines on human labor. Not only did groups such as the transcendentalists and the Luddites raise a furor over this matter, the union movement was in many ways primed by the loss of jobs to automation. The word sabotage allegedly originates from this time, deriving from the practice of throwing sabots, or wooden shoes, into the gears of machines in order to shut them down.

A hundred years down the road and the impact of unions has reached a nadir while automation of manufacturing jobs continues unabated. Robots are routinely employed in assembly lines and have been used for jobs as mundane as drywall installation. Service jobs, those which were seemingly immune to automation, have become the new axis of worker displacement. Even jobs that would seem to require a person are being outsourced to robots with the advent of "Pepper," a "hospitality robot." In Shanghai, robotic waiters are also being employed (Liedtke, 2016). Studies routinely predict between one-third (Manyika et al., 2017) and one-half of all workers (Oesch, 2013) will be eliminated by automation.

It is difficult to pinpoint when the first functions of the service economy began to transform from human to machine operators. Multiple machines—ATM's, toll booths, self-check-out, parking, ticket kiosks in airports, etc.—now perform all manner of interactions that formerly required an individual. The human has been removed from the transaction and all that is required is punching a few numbers or inserting a card and letting the machine take over. In this sense, technological society is decidedly anti-labor.

There is also an element of machine mediated transactions that can be decidedly anti-consumer. This occurs when the machine makes an error. When one is shortchanged by a machine there is little recourse. From the point of view of the machine (anthropomorphizing machines no longer seems too off base) it is infallible. Even when it is broken it is infallible. At this point it might be possible to assault the reader with scenario after scenario but the example of one farmer in North Carolina serves to illustrate this sufficiently. He received a ticket for running a tollbooth in Raleigh, the state capital, while pulling a trailer. Photographs taken by the tollbooth camera led to a match with his trailer which was being pulled behind a Ford pickup. The man in question had not been to Raleigh, did not own a Ford pickup, and the trailer in question had been parked in a barn for six months without wheels. Despite an identification from a grainy photo with poor resolution he was forced to pay the fee in order to renew his driver's license (Cook, 2015).

Automation of services is generally marketed as convenience, but with the exception of an ATM's availability after hours, it is difficult to see how any transactions are more convenient through a machine than a human being. The obvious motivation is financial gain for the provider of the service as opposed to customer convenience. The calculus is simple. When the long-term costs of purchasing and maintaining a machine are less than the labor costs associated with a human employee the machine will get the job. In some cases, the customer is turned into a de facto employee in order for the retailer to replace a human with a machine. The self-check-out line at a grocery store is a prime example of this maneuver. The customer is required to scan and bag their own groceries as well as process payment. This "service" is generally promoted as quicker because the customer does not have to wait in line. Of course, if there were an adequate number of employees and open check-out lines there would not be a wait in the first place. Sadly, many customers seem to prefer the self-checkouts, partially out of a lack of desire to interact with another person. It might be a symptom of the disconnected nature of modern society that there may be a preference for interacting with machines instead of each other.

The removal of workers in favor of machines has costs beyond the loss of jobs. It is also loss of community. The preference exhibited by many for a machine mediated transaction is both a symbolic and tangible loss of human connectivity. The fabric that weaves society together does, at some level, require human interaction. Removing humans from the equation will necessarily have direct and indirect impacts on the very nature of society.

Concerns about the effects of a technologically based society upon workers has existed since the early days of automation. Theodore Roszak, in his book *The Cult of Information* (1986), notes that Norbert Wiener, who coined the term cybernetics in his 1948 text of the same title, and who foresaw the use of "computing machines," *and* who could considered the grandfather of artificial intelligence, was not necessarily a proponent of its development.

> On the contrary, he regarded information technology as a threat to short-term social stability, and possible as a permanent disaster.... Automated machines, Wiener observed, would take over not only assembly line routine, but office routine as well. Cybernetic machinery "plays no favorites between manual labor and white-collar labor." If left wholly in the control of short-sighted, profit-maximizing industrialist, it might well "produce an unemployment situation," in comparison with which ... even the depression of the thirties will seem a pleasant joke [Roszak, 1986].

According to Roszak, Wiener was so concerned about this that he wrote a second book, *The Human Use of Human Beings* (1950), to increase public awareness and encourage discussion of technology in general and artificial intelligence in particular.

Automation in retail also has the potential to manipulate buying patterns. Many stores have systems that allow customers to pay for items the moment they take them off the shelves. Marketed as convenience, the effect is to increase impulse purchases. Similarly, Lowe's Hardware Supply chain is testing customer service robots, called OshBots to help customers find merchandise. Only by a very narrow definition can this be seen as an improvement. What is the problem with a human worker? Is it that the robot does not care about decent working conditions or a living wage?

Many displacements of workers by machines are obvious, but far more occur behind the scenes in such a way that no one notices. There are many features of modern life that have been so seamlessly turned over to machines that we still think humans are in control. An example of this can be seen in the area of broadcasting. Due to regulatory changes, large conglomerates can own multiple media outlets in the same area. Radio stations that appear to be in competition with each other will in fact be housed in the same building where a few engineers run all the stations and carry mostly preprogrammed music. The illusion that music is being selected and played by a DJ is easily created. Only a small minority of the stations are likely to have any type of live programming.

Even activities that did not previously require another human, such as paying bills—or especially paying bills—have become a nearly completely automated online process. Not only do most businesses encourage their customers to do things like pay bills, deposit paychecks or check account balances online, in many cases the individual has no other choice. When customers still retain the ability to pay the old-fashioned way, pressure to switch can be relentless. The outright attempts to manipulate and force one to participate in online bill payments and other types of commerce are constant. Efforts to get the customer to use online payments are often framed in a way that has nothing to do with monetary responsibility. As examples, consider these cases directly from envelopes:

A cable company has on its return envelope—"Spend less time paying your bill and more time doing what you love." This line is followed by another slightly below it and in a smaller font, "Set up easy, automatic bill payments" (Charter Cable). A mortgage company uses this plea: "Switch to secure online statements, they're so convenient and good for the environment" (Wells Fargo).

Leaving aside the argument as to whether online payments are actually "good for the environment," there is the question as to whether paying bills online is easier or just assumed to be? Does it really take any more time to fill out a check and put it in an envelope than it does to go online, login, and make the transfer? The time differential can really be no more than a few seconds if everything works properly. After all, filling out a check and putting it in an envelope can take no more than 30 seconds. How does that compare with the time most people spend getting online on average? Add in the time taken to set up the "convenient" pay method and it is likely a wash at best when it comes to time.

Regardless of what the customer prefers they may be informed that only online payments are accepted. Despite the obvious financial risks involved in conducting online transactions, this appears to be the present and future of paying bills. Banks, more than any other type of organization, should be aware of the security risks involved in online activity, yet these same banks nearly mandate digital banking. Security issues are not the only problems with automatic payments and deposits. Problems related to ingrained cognitive processes are mostly overlooked.

Humans have evolved limits on the ability to process information, the capacity to attend, to remember people and even the mental space available to do all tasks. Certain features of the digital age have an unrecognized effect on cognitive processes in ways that are not fully appreciated. Consider the act of direct deposit. Advantages to these transactions are that the money is instantaneously available (well, let's assume it is, but most banks put a hold on any deposits for at least some time period). This advantage is likely outweighed by one great psychological disadvantage that relates to the need for reinforcement and incentives. Reinforcement works best and is the most satisfying when it is directly connected to the behaviors that results in the reward. This is known as primary reinforcement in the field of learning theory. An example of primary reinforcement would be picking fruit from a tree and eating it. Delaying the reward, using a medium of exchange, and other factors all can diminish the ability to connect the reinforcer to the behavior that produced it. Getting a physical paycheck is already far enough removed from the actions that produced the reward. Direct deposit, however, goes a step further and removes even the tangible representation of reward. Even the representation now is an abstraction. Humans and other animals require tangible feedback for their efforts. Without such feedback, motivation is affected and the connection between effort and reward will be lost.

Conversely, automatic bank drafts also cause unintended consequences. The idea that they free you from the mundane is fine, but they also remove awareness of how much you are spending. Checking in once a month is not the same psychologically as physically spending the money. When a person writes a check to pay a bill it is cognitively processed in such a way that it enters consciousness. Automatic payments lead to financial irresponsibility as the transaction fails to enter conscious awareness.

For many companies keeping customers from thinking about payment has become a strategy. Stored value cards, such as those commonly given as gift cards, are a financial boon for corporations. As soon as the card is purchased money flows to the issuer. However, it is the unused, lost, expired—"spillage" to the industry—portions of these purchases that underlie their use. Spillage is free money. The example of the New York Metropolitan Transit Authority (MTA) will serve to illustrate this effect. The MTA estimates it makes over $50 million a year from unused cards and over $500 million in the last decade (Roberts, 2014). It is further estimated that the total in unused value left on all sorts of cards in 2014 exceeded $1 billion (CEB, 2014). Consumers are apparently becoming more aware of this as spillage has decreased from around 10 percent in 2006 to around 1 percent currently. Most of this change is attributed to the 2009 Card Act that limited the ability of issuers to nullify them (CEB, 2014). However, it should also be noted that these cards have been implicated in significant cases of fraud and money laundering.

Follow the Money: It's About Profit, Not Progress

While the true motivation underlying corporate pushes for online bill payments and stored value cards is clearly monetary, the money involved is small change compared to that involved in spreading the gospel of salvation through technology. Products that were once free, like television and radio, have now become pay services. Formerly the hardware cost money and content was free. Now the hardware is cheap and the content costs.

Monetary incentives compound the pushing of new devices and often force the choice(s) of the user in ways that offer little to no control. The changing of formats (such as in the availability of music) or platforms (incompatible with competitors and even previous versions of the same product) offer little but begrudging compliance from consumers. One must get with the program (pardon the pun) or be left behind. Technological development sees only the end result and the "needs of the user." Interestingly, the "needs of the user" are largely defined by the provider/pusher. Greed, not necessity, is the mother of invention.

To follow this analogy a little further one only has to look to the previously discussed One Laptop Project. Designed to provide cheap, solar powered computers to children in South America it is the joint effort of multiple corporate entities, spearheaded by Facebook, to wire the world.

Ignoring the major issue of cultural disruption, these projects seem laudable in many ways. However, these efforts may also be framed as analogous to a drug dealer building a distribution network and giving away free samples. This is in no way intended to imply bad intent on the part of any of the backers of these projects. They are just providing a service. That is, of course, the rationale of the drug dealer. Free computers to wire the world—it is free crack to get the world hooked.

Maybe a more fitting analogy to describe the digital age lies in the phrase "going viral." Once someone is infected there is nothing that can be done. Computers and connectivity become like insulin, a necessary addiction just to stay functional.

It does not take an astute observer of pop culture to note a proliferation of zombie themed items of interest. From the popular TV show *The Walking Dead*, to the CDC's denial of a "zombie apocalypse," it is hard to escape this meme. It is fair to wonder what exactly is behind this trend.

Lauro (2011) noted that zombies in popular culture appear when there are high levels of societal dissatisfaction. Note a popular band from that other era of deep societal angst—The Zombies—as further support for this hypothesis. The acting out of a mindless, robotic undead indicates psychologically a feeling of being powerless and driven by forces beyond control. A perfect metaphor for the digital age.

Extending this metaphor, it is easy to see zombies as a representation of life being taken away with only a subsistence existence, eating the brains of the living, as a way to survive. However you view it, deliberate infection or addiction, the suggestion is that the outcome is not pretty.

Wearables: The Cyborg Suits Up

While the analogy of infection or addiction is somewhat tongue in cheek, there is more than a small component of this to which there is little choice. To stay connected, modern netizens must become device equipped. Being connected to work and to family and friends while controlling all devices, appliances and everything else on, can begin to feel like a great weight. Laptop, smart phone, iPod, iPad, some type of health monitor, Google Glass, and all the battery packs necessary to power all the devices, leaves the modern person weighed down with connectivity. A look that is decidedly cybernetic in appearance begins to emerge and a new species is easily imagined: *Homo sapiens technologicus*—part human, part technology.

Nearly everyone under a certain age walks around with ear buds; almost everyone of any age carries a phone everywhere they go. Augmented reality goggles are appearing with greater frequency and scores of articles of clothing are manufactured with the goal of accommodating various devices. To be without devices is beginning to be not only unusual, but abnormal for a person living in the modern world.

Despite what appears to be an unstoppable phenomenon there is some push back against wearable technology. A term applied to wearers of Google Glass, "Glassholes," speaks volumes about the receptivity of most to this technology. Some of the pushback, it must be noted, is, as Eric Schmidt proffered, out of a reaction to something appearing new and strange than out of reason. Eventually such technology will become common place and will cease to be noticed. Cell phones used to look out of place, now it is their absence that is conspicuous. Regarding Google Glass or any technology that can record or surreptitiously eavesdrop on behavior, however, it is beyond a fear of novelty that legitimate concerns exist.

Google Glass wearers were subjected to a great deal of animosity when they were initially tested. There were multiple cases of users being assaulted and involved in confrontations (Strange, 2014). It should be noted that in many of these cases the assaults occurred after the users refused requests to turn them off. Several casinos immediately banned their use, more from imagined threats than real ones, although the potential to abuse the technology in casinos seems more than obvious (Costill, 2013, Oreskovic, 2013). In fact, the first wearable computing device was used for exactly the purpose of scamming a casino when Ed Thorp of MIT used a concealed computer to calculate odds in 1961, a practice later banned (PBS, 2016).

In addition to casinos, privacy concerns surrounding Google Glass and other means of ubiquitous recording have been prominent. Citing the ability to clandestinely record and collect vast streams of data, Marc Totenberg of the Electronics Privacy Information Center noted the severity of privacy concerns along with Google's past difficulties with assuring privacy having paid a fine of $7 Million over its mapping project (Oreskovic, 2013).

Responding to these concerns, Eric Schmidt, Executive Chairman of Google has been quoted as saying, "Criticisms are inevitably from people who are afraid of change," and went on to imply most of the negative reaction was related to unfamiliarity with the product (Oreskovic, 2013). Bashing of critics with insults is a time-honored tradition but does not really address the privacy concerns. The idea that technology does not present very real concerns as regards privacy borders on willful ignorance. Part of privacy is the ability to be left alone, to not have to hide. This challenge to privacy in the digital age will be explored more in depth in subsequent chapters. Suffice it to say, privacy advocates will find the digital future an increasing challenge and the future of humans slowing merging with machines seems assured.

Eyeborg

Consider the case of documentary filmmaker Robert Spence, who lost an eye in an accident. He has subsequently had a miniature camera with a radio transmitter placed in the socket. He has proceeded to have a series of conventional prostheses built in order to hide the camera from view (Dark Net: Upgrade, 2016). In this manner he can surreptitiously record constantly without detection. The fact that he continually has the prosthesis modified to produce a more realistic look, and hence a better masking effect, demonstrate his intention to hide his purposes from those he records. The line between human and device is becoming blurred by applications such as this. Of course, it is not necessary to implant a camera to use one.

Body Cameras

As a result of multiple incidents involving police that were captured on mobile devices and public cameras such as the beating of Rodney King, the death of Eric Garner, and others, a move is on to have law enforcement fitted with body cameras. Both sides in the debate appear to advocate for their use. Citizen groups see the cameras as a means to check-up on police while police see them as a means to justify their actions. Multiple municipalities around the nation are beginning to use body cameras as a matter of course.

Research appears to indicate that the presence of the cameras does influence behavior. In a study conducted in Rialto, California, the presence of body cameras was found to have an astounding effect. Use of force by police equipped with cameras was down dramatically, with numbers in the examined period dropping between a minimum of 50 percent to almost 65 percent compared to the three years prior. During the data collection period, in shifts where the police did not have cameras, there were twice as many uses of force incidents as for those with cameras (Ariel, Farrar & Sutherland, 2015; Farrar & Ariel, 2013; Farrar, 2011, Marcelo, 2016).

Citizen complaints also fell dramatically. It was a staggering drop in numbers. Citizen complaints were down between 88 and 94 percent when compared to the three years prior to the study.

Other studies have shown similar effects. In Mesa, Arizona, there was a 60 percent decline in complaints against officers wearing cameras, as compared to the year before, and a 65 percent lower rate of complaints against officers wearing cameras as compared to those not equipped with the cameras (Rankin, 2013; White, 2014; Ready & Young, 2015).

The interpretation of these results has generally been to attribute a "civilizing effect" to the cameras (e.g., White, 2014). Of course, this "civilizing" is really nothing more than an awareness of being watched and recorded. What of the ability to turn off the cameras? This has already proven to be an issue. In an audit of police in Atlanta, it was found that officers commonly violated policy by turning off cameras (Deere, 2018). This seems to be a major limitation of the ability to use the technology to monitor the actions of law enforcement. It would appear that a simple way to prevent abuse of this approach would be to render the cameras not under the control of the officer.

RoboCop (1987), a movie released decades ago, appears more and more prescient as time passes. The story of a murdered policeman who is reconstructed as a machine, it becomes but one more example of science fiction presaging reality. Originally titled *RoboCop: The Future of Law Enforcement,* this film is quickly on the way to becoming all too true.

Law enforcement officers are routinely equipped with all manner of technology from computers to scanners. A well-equipped metro policeman might consistently carry 40+ pounds of equipment between tasers, guns, phones, bullet-proof vests, handcuffs, phones, etc. The modern policeman has not quite yet become RoboCop but the resemblance to Batman's Utility Belt is inescapable. Much of the technology carried by the police, such as phones, tablets, and laptops are also carried by the general public. To exist is to be technologically equipped. For many there is a desire to go well beyond simple wearable technology. For some there is the desire to merge with it.

The Bionic Man

Modern humans may be emotionally stressed but in a physical sense life has never been easier. In this regard the promise of the machine age has been realized. The need to perform manual labor has been drastically reduced to the point that inactivity is more of a problem than overexertion. In addition to removing the need for physical labor, technology is beginning to remove physical limitations of all kinds. ReWalk, a bionic suit that claims to restore mobility to paraplegics, has been demonstrated in early investigations to provide increased mobility with improved health outcomes (Miller, Zimmermann & Herbert, 2016). Using an exoskeleton device, a paralyzed woman, Stacey Kozel, was able to hike the 2000+ mile Appalachian Trail (Crossman, 2016).

Humans have long used technology to adapt to physical infirmities. Magnifying glasses and ear trumpets preceded eyeglasses and hearing aids. Corneal and cochlear implants have been a reality for a number of years. New cortical stimulation methods are restoring sight in those without vision. Biomechanical limb replacement has progressed by leaps and bounds in the past decade. Steve Austin and Jamie Sommers are realities just around the corner.

Steve Austin and Jamie Sommers, played by Lee Majors and Lindsey Wagoner, respectively were the lead characters on the 1970s TV shows *The Six Million Dollar Man* and *The Bionic Woman*. As the result of tragic accidents both were outfitted with advanced bionic technology and became secret government agents. Steve Austin, an astronaut, had been nearly killed when his capsule crashed on re-entry. The voice over to the opening credits intoned: "We can rebuild him. We can make him better."

Embedded in this statement is the belief that machines and technology can improve humanity and are in fact superior to a human. This idea lies at the heart of the newly-termed "extropian" movement.

Extropianism can be loosely defined as the philosophy of the transhumanism movement which seeks to improve humans by technological means. Extropians tend to view humans in a transitory stage and argue for the need to use technology to "accelerate" or push human evolution (The Extropian Principles at www.highexistence.com/the-extropian-principles/). The goals of extropianism appear to be sincere and laudable. However, the implied need to merge humans with machines as exemplified by the term *transhuman* and the initial relationship between extropianism and cybernetics cannot be overlooked. This approach to improving the human condition (i.e., by making us less human and more machine) betrays an attitude that views humans as inferior to machines. Extropianism and transhumanism have at their core a belief that the only way for humans to progress (and extropians believe in "perpetual progress" according to their Manifesto on The Extropian Principles website) is by merging with the machines. This is not just philosophical.

Ray Kurzweil, subject of the documentary *The Transcendent Man* and author of *The Singularity Is Near* (2005), predicts that sometime in the 21st century we will reach the point that humans and machine will merge. He means this not in the sense of prosthetic limbs, implants for motor control, or the like, but rather in the sense that our brains will be "enhanced" by computers. This merger of man and machine is what he means by "singularity." Kurzweil predicts that the pace of change will become so fast that unless we artificially enhance our intelligence the pace of change will become too rapid for normal humans. Applying Moore's Law to human knowledge and intelligence, he arrives at the conclusion that humans will have to be a billion times smarter by 2045 (Kurzweil, 2006) in order to keep up.

This begs the question, Keep up with what? Or who?

The fear that humans will not be able to "keep up" seems to make sense only if the technology is running itself and is in charge. This is an example of refusing to accept natural limits and ruining the present in pursuit of a perfect or ideal future. This notion that humans must change (I hesitate to use the word "evolve" as I think "devolve" might be more apt) in a particular direction is the height of arrogance. Rather than accept that machines can have more raw computing power than our brains, and rather than accept the finite nature of human life, a utopian future is imagined where humanity becomes "enhanced" and merges with the machine. Kurzweil goes even further and anticipates a world where the brain can be uploaded and transferred into a new body, which will be a machine. This would have the effect of allowing one to live forever.

For many futurists this is considered perfection, a utopian state. For others, this is the very vision of a dystopian, totalitarian nightmare where humanity is lost in the name of efficiency. For many of those who ascribe to the techno-uber-culture a path to enlightenment, there is no seeming awareness that for much of the species, humanity is more important than "keeping up." Transhumanists are often very good at looking at the machines but very bad at looking at the humans. There appears to be little understanding of the concept of "too much."

Extropianism is no mere philosophical ambition. The field of biomechatronics, aka bionics, is today a reality. Gone are the clumsy prosthetics that were little advanced from pirates' peg legs. Today bionic limbs are individually designed and controlled using sophisticated algorithms and mathematical models of human movement. Incorporating these advances with those in nerve re-innervation have produced limbs that are more or less under the control of the brain of the user. Much of the funding and research that led to these advances were stimulated by the large numbers of amputees returning from the wars in Afghanistan and Iraq. The historical link between the development of technology and military applications thereof are noteworthy. While funding for basic science is often sorely lacking, the funding for weaponry and the military is seemingly unlimited.

The drawback to the widespread application of much of the bionic technology in development is cost. Unless the tab is being picked up by the military, as through the Veteran's Administration (VA), or some other third-party payer, the price of this technology makes it well beyond the means of the average person. Unfortunately, after soldiers, those most impacted by limb damage are those living in war torn countries or postwar situations, where landmines are either stepped on inadvertently or set off by impoverished people scavenging for metal.

It is in this area that the profit motive and the costs of technology render it useless to the masses of suffering people. The focus in development becomes instead upon perfection of movement and control over limbs. The preference for a high-tech solution tends to remove the economic incentives for more basic functions. Despite market forces to the contrary, all is not lost in the case of prosthetics.

Perhaps the best known, low-tech, functional prosthetic is the Jodhpur limb. Constructed of PVC pipe using basic engineering principles this method provides a low-cost, effective prosthetic. It may not have the ability to interface with the nervous system or to replicate movement exactly but for millions who need a functional limb that increases mobility it is an amazing innovation.

A very exciting development surrounding prosthetics is the production using 3-D printing technology of prosthetic hands for children. Using crowd sourcing methods to design the limbs while simultaneously developing a coordinated network of printers, a group called e-NABLE Community (at http://www.enablingthefuture.org) has linked those with a need for the hand to those who can desktop manufacture them. One of the great features of this effort is that the hands can be made with patterns and colors that look "cool." This has the effect of removing the stigma associated with limb loss in children and creating an opportunity for social status. Further, because of the relatively low costs of the materials, the hands can generally be produced in the $50 to $100 range, comparing favorably to the $40,000+ for a fully innervated limb. With the enormous leaps forward in the associated processes, in addition to hands, 3-D printers and similar technologies hold promise for manufacturing a number of "spare parts" for the human body including already manufactured tracheas (University College, London) and skull and bone implants (Cavendish Implants, England). Everything up to and including organs are regarded as possibilities (Brumfiel, 2013).

Stephen Hawking, in his *Brave New World: Designer Human* (2013), anticipates cybernetic implants to replace failing limbs and parts. He expressed that it might be possible to turn the disabled into superhumans. He also posited that should it become possible, people might then begin to replace/enhance functional body parts in order to gain an advantage. When Oscar Pistorius, the Blade Runner, qualified for the Olympics there were objections that the blades/prosthetics may have given him an advantage. The issue was eventually resolved as basically moot, but Hawking is likely correct.

Steroid use demonstrates the lengths that people will go to gain advantage. If steroids, with their known side effects and almost certain damage, continue to be used, engaging in some form of body modifications seems assured. There are people with a disorder known as apotemnophilia, or Body Integrity Identity Disorder, who feel that they need to have a limb amputated (First, 2005). How long before someone with this condition has their legs amputated, gets bionic replacements and proceeds to break all the high jump records. At what point do we stop being human?

Absurd? Perhaps. Or is Iron Man just around the corner? U.S. Special Operations Command is currently attempting to develop a suit that will make future soldiers nearly invincible. The suit is not only armored but the exoskeleton will be motor driven allowing for increased strength. The device will have a self-contained power supply and provide constant data streams to the wearer. The name chosen for the project?—The "Iron Man Suit," (Cloud, 2013). The mock-up used to present the idea looks so similar to the comic book character that maybe Marvel Comics has a trademark case…

Comic book references aside, none of this is funny. Robot soldiers and drones are already a reality. The concerns over the possibility of terrorists using drones to attack nuclear facilities or other targets is causing great consternation in the intelligence community with "drone repellant" technology being developed as a means to protect sensitive airspaces (Bloomberg Business News, 9/4/16). Is putting a soldier inside the robot for better control, like a Transformer, so far away? Drones are nothing but toys turned into killing machines. Before Rosie the Robot cleans the house, Rosie the Killbot will clean up on the battlefield. A robot has already been used by law enforcement to kill a suspect. In the aftermath of the Dallas police shootings in July 2016 the suspect was trapped, and a robot-delivered bomb was used to kill him (Babwin, 2016). The justification for this use was that it prevented putting officers in harm's way. That is the same justification for robot soldiers. What will be the effect of war being waged by robots? If both sides have the robot soldiers it just seems like an exercise in futility and a giant waste of resources. If it is asymmetrical, with only one side having robots (more aptly, killbots), then it is not a battle, it is just slaughter. There is no place to run, no place to hide.

The very process by which something as laudable as limb replacement is turned into an ethical quagmire illustrates that unpredictable nature of technological development. The unknown outcomes of any effort should be a call for precaution. The consideration of possible ethical conflicts should likely always precede actual development but rarely is that the case. Rather than asking "should?" the question is always "can?" when technology is looking for how to proceed.

This would imply, however, that there is some direction to all of this. One of the lesser discussed and considered issues concerning technological advance is that it largely occurs without conscious consideration of direction or implication. Worries and concerns about unintended consequences are not the province of corporations and techies. Those concerns have largely been viewed as esoteric and left to the philosophers. However, these matters are no longer esoteric. At least three individuals have been fitted with fully functional bionic hands that responded to the commands of their brain (Associated Press, February 26, 2015). Radio frequency (RF) chips have been used for years to track animals and inventory. These chips have recently been implanted in humans to become part of the Internet of Things and can theoretically be used for any number of functions (Dark Web: Upgrade, 2016).

Runaway Trains and Stumbling Behemoths

Development of technological society has progressed in a manner very similar to the way evolution works at the biological level—directionless, purposeless, and unguided. Species, such as the dinosaurs, that at one point ruled the globe, pass into oblivion, leaving only remnants and traces of their existence. The factors that cause extinction, or conversely provide an advantage, are only discernible after the fact. So it is with technology. Just as many individuals falsely see planning and purpose in the evolution of organisms, many also see them in technology. The steam engine and the conversion of coal into energy were the most advanced of technologies two centuries ago; now they are nothing but artifacts of the past. In the case of coal, what is left behind is the imprint of an ancient behemoth, stumbling about leaving a legacy of carbon emissions that will continue to haunt the future long after the last coal fired plant has ceased to operate.

At the time of its initial use, coal-produced energy was the new direction, the new technology that would improve all life. In retrospect, this dinosaur should likely have been killed before it got out of the egg. No one foresaw the ultimate problems that would be created by burning a filthy, heavily particulate substance for fuel. The effect of burning coal was barely noticed at the time of its initial use. This is the way it is with unintended consequences. The implications are not considered at the onset and by the time problems manifest it is too late to place checks on them. In essence, a stumbling behemoth becomes a runaway train.

Might that be what is occurring with technological development? What could be the new coal? Is society building a bigger and more threatening dinosaur while envisioning a brighter tomorrow? Consider the case of the personal computer.

The first desktop computers were little more than glorified word processors. In many ways they were just an advanced electric typewriter with a display monitor. The typewriter, to be sure, required the mining of base metals for parts and likely a coal-backed manufacturing process. Destruction of the natural world and the building of the technological world do indeed go hand in hand. The progression from typewriter to desktop computer to the laptop to the notebook/iPad to the smart phone has not changed this outcome. If anything, it has made it worse.

Out of Sight, Out of Mind:
What You Don't Know Can Hurt You

Production of a typewriter required a manufacturing process that depleted resources and involved the foundry of metals and the exploitation of labor. All of the devices of the digital age also require these same processes. All that has changed is that the effects of the destruction are more widespread. In the days of the typewriter, the devices were made to last a lifetime. Computers on the other hand are on the verge of obsolescence at the time of purchase.

An examination of just the parts that require mining of natural resources brings several facts into alarming focus. One is that a whole class of elements, the rare earth metals, is difficult and costly to obtain. (These metals are not rare; the term was applied based on incorrect assessments and has stuck.) Accurate or not, without these, the digital age would be nearly impossible. Used in a wide variety of devices, these elements make the digital world go. The fact that they are mined under deplorable conditions involving utter environmental destruction is rarely considered. The fact that the end user is far removed from production only adds to the lack of concern as to how they are obtained.

Examination of the source of rare earths provides another source of concern. Monopolies are an emerging fact of life in the digital age. China controls the trade in rare earths metals with over 97 percent of the market share. Despite having large reserves, the U.S. has failed to develop deposits and currently buys *all* of its rare earths from other nations. Control by one nation of such basic necessities creates the potential for both economic and political blackmail. Protesting treatment of dissidents and currency manipulation will never be more than political theater as long as the world is beholden to China for the products necessary to make the digital world work.

Table 7.1. World Production of Rare Earth Elements

	Tons (Metric) Produced	*Market Share*
China	120,000	97.3
India	2700	2.0
Brazil	550	0.42
Malaysia	350	0.27
United States	0	—

SOURCE: United States Department of Interior, Mineral Commodities Summaries of United States Geological Survey (2010).

In addition to the sources listed above, Russia, Canada, Vietnam and Australia have minor reserves. The U.S. has started some minor efforts to develop its deposits since this information first became public. However, the possibility that a crucial element can be completely controlled by any nation should be cause for some reflection.

Another factor that quickly becomes apparent in the discussion of materials such as rare earths and other materials critical to the digital age is that processing these materials leads to enormous environmental degradation along with attendant human costs. In addition to problems such as rising ocean acidity that is destroying shellfish and other food sources, the direct costs to humans are dire. The Blacksmith Institute, which focuses on pollution in developing nations estimates that more than 200,000,000 people are exposed to severe health risks as a result of these actions. The World Health Organization presents more startling claims, suggesting that close to one-quarter (23 percent) of deaths in the developing world are attributable to pollution. These effects are most damaging to children and are known to lead to permanent physical and cognitive effects (WHO, 2009, Blacksmith Institute, 2016).

It should again be noted that those mostly likely to suffer from this type of damage are also those least likely to benefit from the advances of the technological age. As previously reported, toxic chemicals such as flame retardants are present in the bloodstreams of every single individual on the planet, whether they live in a technologically advanced society or not (Cribb, 2014). The magnitude of this type of problem, worldwide effects of pollution, is only just beginning to enter the consciousness of the world.

Exposure to toxins and their byproducts, substances necessary to maintain the technological direction of the world society, is unavoidable. Recent studies have also suggested that air and water pollution are having major effects. Studies have tied pollution to increases in the rates of autism (Beccera, Wilhelm, Olsen, Cockburn & Ritz, 2013; Dawson, 2013; Vol, Lurmann, Penfold, Hertz-Picciotto & McConnell, 2013; Seaman, 2012; Volk, Hertz-Picciotto, Delviche, Lurmann & McConnell, 2011; Herbert, 2010), dementia (Adams, 2017; Chen et al., 2017), Parkinson's, multiple sclerosis (Chen et al., 2017), and respiratory illness (Seaton, Godden, MacNee & Donaldson, 1995) among other problems. The effects of chemicals in the water supply from such actions as pharmaceuticals poured and pissed in the drains are also being recognized (Cribb, 2014).

Unfortunately, the problems of toxic substances in the environment go well beyond air pollution and climate change. Technological advance, largely unplanned, but motivated by economic incentives, has led to the development of an untold number of products that may have life altering or even life-threatening effects. Monetary realities suggest that even when the effects are known there is unlikely to be a cessation of production or use. These products are largely composed of, or coated with, industrial chemicals, and are unavoidable by anyone alive today. It goes beyond mere exposure to pollution or even flame retardants. Everywhere, everything and everyone is touched by the residue of the technologically advanced society. To live in the current time is to be bathed in chemicals.

Bathed in Chemicals

The use of chemical compounds in manufacturing and industry is poorly understood by the public at large. Most people have a false sense of complacency that envisions a complex layer of government regulations and regulators that protect the consumer and prevent dangerous products from reaching the market. This belief is incorrect. The vast majority of products have absolutely no testing conducted on them whatsoever. The Underwriters Laboratory (UL) is a private organization that barely scratches the surface of the products brought to market. Aside from UL, almost nothing gets tested without a grant. Individual compounds, component parts and the like, have scant oversight. In fact, not only are most chemical processes not independently tested, they are not even publicly known.

In industry, the development of new products and processes involve proprietary information and as a result are largely kept secret. New chemicals and processes are neither patented nor revealed out of fear that rivals will benefit from them should they become known.

Rather than following the precautionary principle, what happens in the free market economy with protection of proprietary information and an emphasis on innovation, is a situation where new toxins and poisons can readily be introduced. Until a chemical produces an obvious effect, i.e., an effect so strong that it happens to be noticed even when no one is looking, there is essentially *no testing* of industrial material for toxic or health effects. Rather than a precautionary principle the free market operates on a "prove it's bad" principle. Think about how long it took to "prove" cigarettes were dangerous. What should have been learned from cigarettes is that industry cannot police itself. Money, especially lots of it, can make it very hard to do the right thing, or even to not do the wrong thing. In the case of the cigarette companies it is now know that the companies covered up the research they had showing the dangers of smoking.

As go the cigarette companies, so go the chemical manufacturers. For most chemicals used in industry there is absolutely no system of checks other than the manufacturers themselves. Chemicals are essentially still regulated under the outdated 1976 Toxic Substances Control Act (TSCA).

This act does not require chemical compounds to undergo any sort of testing prior to market introduction nor does it even require these items to be registered. The act grandfathered in 62,000 compounds in 1976 on the rationale that they had not caused any obvious damage up until then. As such the TSCA only applies to new substances.

While companies intending to manufacture a new chemical must notify the EPA, safety information and toxicity testing is not required before release. In fact, TSCA requires the EPA to determine that the chemical will "present an unreasonable risk to human health or to the environment" before it can even begin to regulate a chemical (TSCA, 1976). This hurdle of "unreasonable risk" makes regulation nearly impossible. The risks of many chemicals may not be readily apparent. Lead has been used for millennia, but it has only been within the past fifty or so years that humans figured out its very potent effects. Political ineptitude and effective lobbying efforts on the part of industry have essentially prevented meaningful change and updating of this TSCA. Only in early 2016 did Congress pass an update of the TSCA and begin to regulate at least some chemicals such as BPA, formaldehyde, asbestos and styrene (Daly, 2016). Unfortunately, recent efforts by the EPA have begun to undo even some of these minimal regulations (McCall, 2018).

Estimates of the number of chemical compounds in existence are difficult to determine with a high degree of accuracy. In addition to the 62,000 chemicals grandfathered in 1976, there have been estimated to be at least 50,000 (Fischetti, 2010) to as many as 85,000 (Urbina, 2013) other chemical compounds in existence. Because of matters such as industrial secrets the number is surely even higher. Of all these chemical compounds the EPA tracks 660 (source: Union of Concerned Scientists, 2019) and has tested only about 300 of these, with restrictions placed on only five of those (Fischetti, 2010).

The chemicals used in the fracking process highlight the way such compounds are actually treated. Is fracking dangerous in terms of its effects upon the water supply? The answer is not really known because the companies will not reveal what exactly is in the mixtures used to force out the gas, holding it to be a proprietary secret. The real effect of this is to prevent any independent analysis.

The time taken for the peer review process in research and the time taken for judicial review only add to the problems of regulating potentially dangerous chemicals. Compound this with the myriad of ways to tie things up in courts, such as patent disputes, and this ultimately leads to the failure of current methods to prevent dangerous chemicals and materials from coming to market.

The effects of these scientific and judicial constraints do not serve to protect the public but rather have the opposite effect. The deliberate concealment of information leads to a lack of oversight and resultant health consequences for everyone. These chemicals fall into many categories and are ubiquitous. They cannot be escaped as they are everywhere. Lack of knowledge of their existence clearly makes evaluation of their potentially damaging effects nearly impossible. Even when problems with certain chemicals are suspected or even known to exist it may take decades before the dangers manifest at such an obvious level as to be fully evaluated. At that point, the damage has been done.

A Few (Very) Little Things to Worry About

At some level it is impossible to enumerate all the chemicals that are both the byproducts and the fuels for the modern era. The amount of toxins is simply too great to discuss in any breadth. However, a few categories of microparticles are of particular note. Some of these are:

MICROBEADS: Used in cosmetics, etc. These beads float on water and water treatment plants are unable to remove them. They have spread throughout the Great Lakes region with high concentrations found in many areas. Research conducted by Loyola University found 1,500 to 1,000,000 microbeads per square mile (McCormack, Hoellein, Mason, Schleup & Kelly, 2014). Other studies have demonstrated the ubiquity of this type of trash in the environment (Drisel et al., 2015; McCormick & Hoellein, 2016). The state of Illinois initiated a production ban beginning in 2019.

MICROFIBERS/MICROPLASTICS: The Florida Microplastics Awareness Project (FMAP) has found that the presence of strands of plastic fibers poses one of the most serious problems as regards ocean pollution. Finding plastic in 89 percent of all samples tested, with over 80 percent containing fibers of plastic residue, the presence of this difficult to degrade substance in the environment is causing problems for wildlife and humans (FMAP, 2017).

TURF: Artificial turf has been found to contain arsenic, benzene, nickel, cadmium. In an anecdotal report from an oncologist, 38 soccer players, 34 of whom were goalies, were found to have lymph related cancers. The high proportion of goalies treated was attributed to the fact that they spend more time in direct contact with the ground (Morning Joe, 2014). High lead levels have been found in at least one study of turf (Pavilonis, Weisel, Buckley & Lioy, 2014).

TEFLON: Perfluorooctanoic acid (PFOA). The EPA fined its producer, Dupont, $16.5 million in 2005 for deliberately covering up health problems including birth defects and cancer (Savan, 2007). Production of this product has ceased due to its toxicity and permanence. Unfortunately, Teflon remains in the environment forever. Detected in the bloodstreams of dolphins, polar bears, and almost all people, it will continue to be with humanity in perpetuity.

GRAPHENE: Graphene Oxide. Clings to organic matter. The primary effects are on surface water (Bourzac, 2014) where it floats around and becomes stuck to organic material (Walker, 2014).

PHARMACEUTICALS: Disposals of pharmaceutical products of various types are leading to an infection of the water supply. As it turns out, people routinely dump leftover pharmaceuticals down the drain. These particles are not generally susceptible to the effects of water treatment and as a result remain within the system. While there is a tremendous dilution, the levels of these chemicals remain high enough that they have been detected in several studies (Heberer, 2002, Webb, Ternes, Gibert & Olejniczak, 2003; Jones, Lester & Voulvoulis, 2005). The long term effects of unintentional medication through the water supply are not even remotely known.

NANOWORLD: The nanoworld is one that is shrouded in mystery for most individuals. The problem with the nanoworld is exactly that which makes it useful—i.e., its size. There is concern that the nanoworld might "escape" and begin to "infect" other organisms. The ability to calculate the probability of such events is in the realm of speculation.

PATHOGENS: The broad array of pathogens that are known to humanity have become potentially more lethal since the advent of air travel. Only within the past 100 years has air travel become a factor in the spread of disease. It is only within recent decades that the possibility of a worldwide pandemic as a result of air travel has emerged as one of human concern. The more recent past has seen the emergence of groups willing to use biological weapons on a large scale to wreak havoc. Technology is not the cause of pathogens, but it can be a means of increasing exposure in ways not previously possible.

Curiously, the lack of concern exhibited over the presence of untested industrial chemicals in the environment is accompanied by what might be seen as an over concern with germs and bacteria that exist in nature. Concern with germs or contact with dirt are absent among the other members of the animal kingdom. Among humans, in an expression of alienation from nature, the organic world is viewed as threatening. The presence of germs is seen as antithetical to a sterile technological existence. Toxic cleaners are applied in industrial settings to "sterilize" the field. Even in daily life there is something of an obsession with keeping germs at bay. The ultimate expression of this can be seen in the use of portable bottles of cleaning agents and hand sanitizers to kill germs. These substances do kill germs, it is true. Unfortunately, it appears that they kill germs too well.

It is beginning to be noted and well documented that an environment completely devoid of germs may not have the health effects implied but instead may actually have a deleterious effect. Immune systems, particularly those of children, become impaired or worse, fail to develop properly, without some stimulus. In a study the immune systems of children raised in exceptionally clean, nearly sterile environments failed to develop properly. Children raised in less hygienic conditions who spent time playing with dirt and being exposed to germs had more vigorous immune systems (Olszak et al., 2012). In another study conducted subsequent to the fall of the Berlin Wall, children raised in East Germany and exposed to higher levels of germs exhibited stronger immune systems than those raised in West Germany (von Mutius, 2007; 2010). The lead researcher in this study, Erika von Mutius, who is also one of the world's leading researchers of asthma, has proposed what has become known as the "hygiene hypothesis," which posits that healthy immune and respiratory systems need certain bacteria introduced into the system in order to encourage proper development and response of these systems. Investigation into this hypothesis from academic researchers demonstrates solid evidence for this concept (Bach, 2018; deLaval & Sieweke, 2017; Lambrecht & Hammand, 2017; Garn & Renz, 2007). A way to conceptualize this response is to think of it as similar to the way vaccines work. Recent investigations into peanut allergies appear to provide further confirmation of this hypothesis (Du Toit et al., 2015). In sum, evidence suggests that an environment can be too sterile for the proper stimulation of the immune system.

Compounding the problem of impaired immune systems is the secondary effect that has been noticed with modern germ-killing agents known as triclocarbans. These chemicals are used in hand cleaners and are complicit in the rise of "superbugs" as antibiotic resistant infections are sometimes called. In fact, it appears that the more efforts given to killing germs the greater the likelihood of producing superbugs.

While these sanitizers are extremely efficient at eliminating common germs (99.99 percent is the claim) the germs that are left behind can be considered "super germs." These germs tend to be particularly resistant to any efforts to eliminate, and when not forced to compete with common germs, multiply at higher rates. This leads to conditions in which germs that could be handled by the immune system are absent while those that are particularly difficult to control proliferate. Multiply Resistant Staff Infections (MRSA) and other nearly untreatable diseases are resultant. The U.S.

Food and Drug Administration (FDA) has specifically implicated antibacterial soaps in the problem and banned them as unsafe (Tavernise, 2016).

Hand cleaners are not the only culprits in this problem. Other efforts to control the spread of disease and infection have backfired to a large degree. According to the Centers for Disease Control it is estimated that 23,000 die and more than two million people are afflicted every year as a result of antibiotic resistant infections. The non-judicious use of antibiotics in both medicine and corporate farming are also important factors in the proliferation of drug resistant diseases and infections. Livestock uses up to 80 percent of all antibiotics and as such are a major factor in this problem according to the World Health Organization (2013). The end result of all this overuse and misuse of antibiotics is a lessening of their effectiveness. Tom Frieden, head of the CDC, has said "Our strongest antibiotics don't work…" and a review of hospitals found antibiotic resistant bacteria in 3.9 percent of all hospitals, in 17.8 percent of specialty hospitals, and in 42 states (Begley, 2013).

Antibiotics, once considered nothing short of a medical wonder, are being rendered useless (Blaser, 2014). Decades of warnings concerning this problem have done little to stem the overuse of these drugs (Epps & Walker, 2006; Harrison & Svec, 1998).

Corporate farming, as noted, contributes to this problem. Livestock are routinely administered whopping doses of antibiotics and other drugs as standard course. Other substances such as hormones that add weight to animals are also being implicated in health problems. That which is put into animals and plants to make them grow faster is then transferred to the consumer of these food products. Certain chemicals, particularly endocrine disruptors, appear to have had a significant effect and have been identified as a causal factor in the decrease in the age of menarche (Aksglaedem et al., 2006; Blanck et al., 2000; Colon, Caro, Bourdory & Rosario, 2000). Other studies have linked hormones in the food supply, especially milk, to metabolic problems. This occurs as a result of cows that are administered recombinant Bovine Growth Hormone (rBGH)—they secrete an insulin-like metabolite that gets into the milk (Parent et al. 2003).

Multiple health problems are being attributed to the methods used in factory/corporate farming. In an investigative report by *Mother Jones* magazine, "Gagged by Big Ag," the hidden nature of the process of corporate farming is detailed (Genoways, 2013). While this report focuses primarily upon the lengths to which the corporate food industry goes to shield its practices from scrutiny, the secondary feature of the report relates to the means by which factory farming sacrifices the health of the consumer in the name of profits.

Most consumers have little understanding of the actual way food is produced. Visions of small family farms are largely illusory with corporate entities largely controlling the manufacture and protection of food. The degree to which the desires of consumers are ignored in the production of food is but another reflection of life in the machine.

Genetically modified food is one example of the way a technology with but moderate benefit, perhaps with the exception of profit for the producer, is forced upon consumers. Genetically modified food is widely distrusted and not preferred by consumers. Arguments against the validity of these concerns aside, there is clearly a preference against GMOs in the food chain. Many European nations require labeling, while in other places, notably India, there are violent uprisings directed at GMO crops. Lobbying by industry in the U.S. has largely prevented labeling, even labeling that states a food is GMO-free, to such a degree that knowledge of the issue is lacking. The extent to which GMO crops exist is not common knowledge.

Table 7.2. Increase in Genetically Engineered Crops Over Time (in millions)

	Corn		*Soybeans*		*Cotton*	
Year	*Acres*	*% of Crop*	*Acres*	*% of Crop*	*Acres*	*% of Crop*
2000	19.89	25	40.10	54	9.47	61
2005	42.53	52	62.67	87	11.25	79
2010	75.85	86	71.99	93	10.21	93
2013	87.64	90	72.29	93	9.23	90

SOURCE: Fernando-Cornejo, J., Wechsler, S., Livingston, M., & Mitchell, L. (2014). Genetically engineered crops in the United States. February 2014. United States Department of Agriculture. At www.ers.usda.gov/media/1282246/err162.pdf

It was further estimated by the USDA that farmers planted approximately 170 million acres of genetically modified/engineered crops in 2013, which was about half of all land available for planting (Fernando et al., 2014). Corporate entities exhibited a disproportionate share of control over the application of GMOs.

Table 7.3. Approved Releases of Genetically Engineered (GE) Crops by the USDA

Entity	*Number of GE Crops*
Monsanto	6782
DuPont/Pioneer	1405
Syngenta	565
Dow Chemical	400

Entity	Number of GE Crops
USDA Agricultural Research Station	370
AgroEvo	326
ArborGen	311
Bayer	260
Seminis	210

SOURCE: United States Department of Agriculture—Animal and Plant Health Inspection Service (APHIS)

What is generally buried in discussions concerning GMO food crops is the primary reason behind the modification. The assumed and often promoted reasons for GMO's are that they increase crop yields, although evidence for this claim is mixed at best. The factor that lies at the root of most GMO's is that they are modified to withstand pesticide applications. Although the general public tends to worry about the effect of modifying the food the real concern should be the presence of the pesticides during growth and the residue thereof that remains on food. In many ways the debate is about the wrong concerns. It may be that it is not so much how the genes for the food have been manipulated but rather the reasons for the manipulation in the first place.

Previously banned pesticides demonstrate the potential they have to affect human life. After the pesticides diazinon and chlorpyrifos were banned, newborns increased in size in areas where the chemicals were used. Before the ban the size of newborns could be correlated with the levels of insecticides found in the blood of mothers and in umbilical cords (Whyatt, 2004).

The inability to resist the influence of corporate agricultural interests goes beyond simple lack of knowledge of what is in the food supply. Even the ability to control the food supply and keep corporate interests from forcing their will upon all of agriculture is currently at stake. In Bowman v. Monsanto Corporation (SCOTUS #11–796) a farmer was held to have violated the patent of Monsanto over genetically modified soybeans. Mr. Bowman purchased the soybeans from a grain elevator where they were unlabeled and advertised as a mixture of soybeans from various sources. Monsanto prevailed 9–0 in the court case. The Patent Exhaustion Doctrine was the basis of the legal ruling. This law basically states that the buyer of a patented object can buy or sell it but not make copies of it. This ruling, applied to "Self-Replicating Technology," essentially gives corporate entities control over the food supply. Buried in the ruling was the idea that Bowman should have known there were Monsanto varieties in any "mixed" group because of the fact that such a high percentage of the market was controlled by them (SCOTUS #11–796).

In a separate case, Organic Seed Growers & Trade Association et al., v. Monsanto Company et al. (SCOTUS #13–303) lost a lawsuit against Monsanto after having their fields ruined by contamination from genetically engineered crops nearby. Canadian alfalfa farmers Percy and Louise Schmeisers had spent fifty years developing strains of the crop only to have their fields cross-pollinated by those using Monsanto varieties. Monsanto aggressively went after the couple, not only ruining their livelihood but demanding damages as well. The Schmeisers lost (The Grand Illusion, 2015).

This legal imprimatur of corporate control over the food chain is of relevance far beyond the destruction of family farms. Corporate practices related to animal husbandry have long been implicated in matters such as antibiotic resistance. Perhaps the greatest underlying problem with large corporate control of every aspects of the food supply has to do with concomitant standardization and regularization of food. Diversity of genetics in crops and animals is restricted as a means to produce a uniform product. Unfortunately, a necessary effect of eliminating diversity is that the food source then becomes vulnerable to diseases and other problems. A blight that affects strawberries might only hit 10 percent of varieties, but if all varieties but one cease to exist, anything that affects that variety could prove disastrous. Bananas are currently so threatened due to corporate farming practices on banana plantations (Koeppel, 2008; 2011).

These rulings also have unfortunate implications for information. Essentially it allows for the commoditization of information in the guise of protecting the copyright holder. The fact that a biological organism can be patented, though genetically engineered, is a slippery slope. Treating biological organisms as if they are machines akin to a can opener is an issue that must be a matter of serious debate. It is another step down the road to equate the organic and the technological. This equation should not compute. Biological entities are not machines.

However, the lines between biological entity and technological application are becoming blurred. All manner of intrusive technology is accepted as normative and deemed necessary for existence. Humans are coming to feel they must augment their biological being with multiple devices in order to stay even. There is little choice but to accept the technological overlay that permeates society. In some cases, the restriction or alteration of behavior may be little more than an inconvenience as delineated in the early portions of this chapter. However, the insidious and often hidden nature of the technological overreach that is intertwined with all of life can also lead to severe unintended effects ranging from a planet poisoned by toxins, declining physical and mental health, and a compromised food supply all in the service to convenience.

8

Big Data, the Surveillance State and the Death of Privacy

I want to be let alone.—Greta Garbo

The technotronic era involves the gradual appearance of a more controlled society. Such a society would be dominated by an elite, unrestrained by traditional values. Soon it will be possible to assert almost continuous surveillance over every citizen and maintain up-to-date complete files containing even the most personal information about the citizen. These files will be subject to instantaneous retrieval by the authorities.
—Zbigniew Brzezinski, *Between Two Ages: America's Role in the Technotronic Era*

BIG DATA

In the late 1920s and early 1930s a mathematical construct emerged that has come to be known as "Six Degrees of Separation." The idea is generally attributed to Hungarian novelist Frigyes Karinthy (1929), although others have stated the idea originated with Marconi. It is a theorem that hypothesizes that any two people can be linked by a string of six or fewer connectors. This theorem was experimentally confirmed by Stanley Milgram (1967) using post cards addressed to individuals unknown to the subjects who were instructed to send them to someone who could assist in forwarding the card to the addressee. Milgram found the average number of mailings to be 5.7 for the card to reach its intended target.

This work has been replicated in a number of investigations, most notably at Columbia University (Dodds, Muhamad, and Watts, 2003). A theoretical mathematics presentation of this theory has been available online for years (de Sola Pool & Kochen, 1978). This theorem is the basis for social network theory and is the underlying basis for many meta-data analyses.

This concept of six degrees of separation has become a familiar meme in popular culture, inspiring a movie, a television show, and even an Internet game: Oracle of Bacon, aka Six Degrees of Kevin Bacon. Albert-Laszlo Barabasi in his seminal text *Linked* (2003) demonstrated how this model extends beyond theoretical mathematics and into the real world, connecting everyone in every possible way. Utilizing algorithms, basic information is turned into a numerical code that builds upon this theorem.

> **al·go·rithm** (ăl´gə-rĭth´əm) *noun*. 1. A step-by-step problem-solving procedure, especially an established, recursive computational procedure for solving a problem in a finite number of steps. *The American Heritage Dictionary*, 4th ed. (Houghton Mifflin Company, 2002).

Without getting into the mathematical weeds on this issue, algorithmic procedures serve as a means to take the various numeric codes generated by users and proceed to sort, match and catalog them. The ability to connect people, then to extend these connections into a means of categorizing individuals to determine their social networks, consumer preferences and any other factor that can be imagined, is the essence of meta-data analysis and the use of this has come to be known as BIG DATA.

Big Data, the name given to the use and manipulation of massive data sets that can be used to predict behavior, commerce patterns and extremely personal information, is ubiquitous (behind the scenes) in the cyber world. Number crunching and statistical analysis devoid of emotional interference, the process is nothing more than applied mathematics. Using aggregate data to predict the activity and behavior of individuals essentially become nodes on a network. An excellent treatise on this concept is presented in *Big Data: A Revolution That Will Transform How We Live, Work, and Think* (Mayer-Schonberger and Cukier, 2013).

Big Data has been found to have the ability to predict outbreaks of communicable diseases and provide assistance in the preparation for epidemics such as influenza (which it successfully did in 2012 and 2013). When correct, this use of data is wonderful and extremely useful.

In addition to the prediction of disease Big Data also has a number of applied uses. Advances in computing, specifically as regards the ability to do more than store information, with capabilities that allow computers to manipulate information and create networks through the use of algorithms, have led to a number of extensions. Predictive programs have been applied in areas as diverse as music selection to policing.

Twenty years ago, a joke about computers was that their biggest problem was that they did exactly what they were told. The implication being that the user was ultimately responsible for any problems in operation. Today computer programs have turned 180 degrees. Some devices have become so complex that they are beyond the mere process of input leads to output. Modification of information, feedback based on previous activity, interaction with other devices, whether intentional or accidental, all are beginning to change the very fabric of life. To find a restaurant, just state what type of food is desired and the phone will immediately find multiple options and plot out the course to get there. Break a tooth while eating, emergency dental services are just a smart phone search away.

Ironically, this ability of the machines to predict what one wants may prevent serendipitous discovery. Instead of lives being enhanced, everyone is being pigeon-holed into neat little algorithmically determined boxes. A box, inside the machine.

Technological innovation often occurs with little concern for the unintended consequences. Data used to find friends on social media can be the same meta-data used to infer guilt by association. This guilt by association approach, when used as a loophole by security agencies, becomes the means by which everyone can be legally swept up into surveillance. The ACLU has provided a trenchant example of how this can be used as a loophole to spy on everyone, by the case of a terrorist calling a pizza delivery place. By linking the suspect/target of surveillance to the phone number of the store, the FISA law then allows the collection of the data of everyone else who calls the same number (STAND, 2015). A behavior as innocent as ordering pizza becomes a pretext for surveillance. This loophole in the FISA law clearly creates a situation where everyone is a suspect. If the NSA were to consider everyone "connected" to a terrorist as anyone within six steps of connectivity it is almost mathematically certain that the whole planet can be swept up into the web of data collection.

This is justified as necessary because of security concerns. When governments that allegedly stand for freedom bend the law and find loopholes to get at the data, can there be any doubt that repressive governments will do the same? Oppressive governments are going to err on the "safe" side against dissidents. Totalitarians and dictators rarely concern themselves with matters of privacy. Safety is a justification for oppression. Monitoring the Internet is not the sole province of the U.S. national security apparatus. For nations such as North Korea or China, the internet, aided by data analysis, becomes another means for spying on citizens. With knowledge of meta-data, entire networks of dissidents can be monitored and identified. Social media may provide a message post for revolutionaries, but it also provides a tracking mechanism for the oppressors. Some people do not get this.

Transparency Movement

Who started this? The government of North Korea, China, Iran and/ or the NSA?

Presented as an effort to make sure that the Internet is secure, and that people are "who they say they are," the transparency movement is at best a misplaced effort and at worst a potentially life-threatening blunder. Mostly portrayed as a means to keep in check the "bad actors," the scammers, slammers and thieves who haunt the shadows of the Internet, this movement either conveniently overlooks or attempts to deny the fact that anonymity is one of the primary vehicles that moves against oppressive governments. Google and Facebook have made concerted efforts to get their customers to use real names or "offline identities." China also requires its citizens to do the same.

It is easy for people living in the United States or Western Europe to say that everyone who has an opinion should willingly have their identity attached to it. This effort has more to do with outmoded ideas of "homeland security" and efforts to catch people than to "protect" anyone.

A case often presented by the transparency movement to justify its approach is that the founders of the U.S. proudly proclaimed their grievances and signed the Declaration of Independence publicly. This makes a case that all political dissent must occur in public. It denies the reality that the Boston Tea Party was a group of Freemasons dressed up as Native Americans in order to hide their identity. *The Federalists Papers* were originally published as by "Publius," and Ben Franklin wrote as "Poor Richard." Let us not forget François-Marie Arouet, better known as Voltaire, who argued that anyone with anything controversial to say should hide their identity. It is naïve to the point of disingenuous to suggest that political dissent must be public and identifiable. This might be an easy position to take for Western Europeans or Americans on Wall Street. It is not so easy to take in front of a tank in Tiananmen Square or on the streets of Kabul. The tracking of dissidents, under the guise of transparency, is the dream of oppressive states. The iPhone had to issue emergency upgrades after it was revealed that the phone could be accessed with the tap of a finger. This flaw went undetected until rights activist Ahmed Mansoor realized he was being targeted (Satter & Chetlow, 2016).

For those who can rationalize their own government's spying on them in the name of national security or even corporations' using the data to market to them, consider the case of foreign government intervention. Specifically, China has been involved in coordinated and far flung hacking efforts. What appears unique about the Chinese attacks is that government activities are not its sole target. The attacks upon corporate interests have gone largely undetected with a few exceptions.

In one instance it was found that certain smart phones were sending all meta-data to an address in China every 72 hours. The software makers said that the code, specifically written to carry out the function, was a "mistake." Responsibility for this action has yet to be determined (Apuzzo, 2016). In June 2014, five Chinese military officials were indicted by the U.S. over cyber-espionage, much of it directed at corporate officials according to a statement issued by the United States Department of Justice. The officers, working out of Military Unit 61398, were engaged in numerous activities and corporate sabotage. China insisted that all of its activities were legal and immediately accused the U.S. of spying (Elias, 2015).

A more sinister version of potential state-to-state cyber espionage came to light in November 2014 with the revelations by Homeland Security officials that a computer virus dubbed "Black Energy," appears to have been placed on an extremely high number of public utility plants throughout the United States. This virus, apparently of the simple Trojan horse variety, has been found on the systems of water and power plants. The virus gives those responsible access in a manner that will allow them to overload systems, flood water plants, cut off electricity, manipulate the grid, and perform numerous other actions (Cloherty & Thomas, 2014). The U.S. government issued an alert regarding this situation to all critical infrastructure sectors designating the Russians as the primary suspect (U.S. CERT Alert, TA18–074A). The Russians are widely suspected of using the Black Energy, or Sandworm, virus to disrupt power in the Ukraine on two separate occasions.

The Russian government has also emerged as one of the primary players in state sponsored hacking. Its efforts interfering in elections in the United States are of enormous proportion, revealing the coordinated actions of at least two different hacking groups under control of the state apparatus (Mueller, 2019). Their efforts are ongoing to interfere in U.S. society through a wide variety of actions from social media campaigns to espionage.

Of course, governments' spying on each other is nothing new. Espionage is part of national security. All sides lament what the others are doing but also plead necessity. The ability and seeming will to attack vital infrastructure set the attacks by Black Energy in a category different from most state spying which is usually directed at obtaining information. The computer virus moves into the direction of cyber warfare and espionage as the new normal. From almost every perspective the possibility to use technology for invasive purposes is a new frontier. All sides will surely find rationales to use any means at their disposal. Although never conclusively proven, the Stuxnet virus that damaged Iranian centrifuges is widely believed to have originated from within U.S. or Israeli governments (Kushner, 2013).

The activities of Russia and other national governments brings the discussion back to the underlying rationale of all surveillance and data collection programs—to make the nation and world more secure. There is scant evidence that all of this data collection and questionable surveillance has made the world safer. In fact, there is counter evidence to suggest that the collection of so much information has created the circumstance whereby the data cannot be properly sorted and assessed. Terrorism continues, and while Bin Laden and numerous other terrorists have been killed, bombs continue to go off all across the world and the rash of mass public shootings in the U.S. continues unabated. What seems to have been accomplished is a false sense of security.

Chinese government hacking also raises the specter of corporation-to-corporation spying. Industrial and corporate espionage are nothing new, but in the present time frame it is part of doing business. Because the law is often so murky or even silent on these issues in many jurisdictions, some corporate spying may not even be illegal.

Also, the repression, intentional or not, of artistic expression through the guise of making everything transparent, cannot be overlooked. Artists often use pseudonyms. Sometimes the use of pseudonyms is important to the overall presentation. Performing under an assumed name might seem nonsensical to a lawyer or banker, but to many artists it is an integral part of their presentation, e.g., Blue Man Group or Daft Punk. How far does transparency go? Must Bono have "Matthew Hewson" beside his name at all times? Would Mark Twain have been allowed a webpage under that name or would he have to register as "Samuel Clemens"? It would likely be argued that because these two are famous they would of course be allowed to use their aliases. But how did they get famous? Every person who is famous under a persona was at some point not famous under the same persona. How many people would know that Robert Zimmerman is Bob Dylan? How does this get resolved with transparency?

Big Data is about far more than covert government surveillance. A very significant use of data sets does not concern violations of privacy per se, but rather is related to the gathering of information for the purposes of commerce. This is what is really driving the transparency movement—the desire to identify potential customers for marketing purposes. In other words, it is not just the NSA that uses meta-data; it is also business, big and small, as it begins to micro-target and micro-market to consumers. The vast majority of individuals, many of whom are infuriated by the government's tracking their online activities for security purposes, willingly give away vast quantities of personal information every time they enter a commercial website.

Giving It All Away:
Big Data and the Corporate World

It is not only from malevolent cyber-spammers and hackers bent on stealing information through any means possible that it is necessary to be concerned with cyber security. Most individuals are their own biggest threat. The casual nature by which most people approach their cyber activities is alarming. Young people, who should be savvy, appear in some ways the most susceptible. Having become comfortable in the digital age, little thought appears to be given to what one posts. In order to be connected and create a digital presence many people are unknowingly, and often knowingly, giving away their information.

A common Internet joke is that the biggest lie is "I have read and understand the terms of service." Who possibly could? Most of these are written in legalese and lengthy enough to ensure that they are rarely if ever read. Almost all terms of service agreements are presented on a take it or leave it basis. Many of the outright invasions of privacy described herein are usually covered by a loophole or some vague disclaimer that is buried deep in the terms of service agreement. In the physical world this might be akin to a key to your house being sent to City Hall or the local police upon purchase.

The modern world is the paranoid's fears come to life. There exists the ability to track almost everyone all the time. Some people see this as a good thing. Children tracked by parents for safety reasons, people on probation electronically monitored by the courts, programs and apps that allow people to find friends. Some corporations are implanting radio frequency (RF) chips in employees (Progressive, 2017). All are viewed as social goods by some segment of the population. To others it is like life in a dystopian novel. Everything is tracked, everything is recorded. Just in case it is needed. An instrument that can be used to disseminate thought is instead being used to limit it. Wear the chip and don't complain. What could go wrong?

Police efforts using technology are already beginning to exhibit the overreach that comes with technological advance. A program called "Sting-Ray" that mimics cell phone towers to collect meta-data and is already in use along with similar/complementary programs called "Amberjack," which tracks and locates cell phones, and "Kingfish," which collects identity codes and makes connections between phone users (Gonzales, 2014, June). The ACLU has noted that police are systematically hiding information about these programs, which it has identified are employed by at least a half-dozen law enforcement agencies, including Homeland Security, and more than 40 police departments (ACLU Website). This number is almost assuredly much greater as investigative reports have discovered that law enforcement agencies have been hiding their use of such programs in the guise of national security, going so far as to drop cases to prevent knowledge of the systems from becoming public (Gillum & Sullivan, 2014).

Perhaps more alarming is what appears to be a growing private-public exchange involving private entities and law enforcement that views private data as a commodity and law enforcement as customers. Investigative reports have exposed that the Digital Recognition Network and the National Vehicle Location Service (sounds official doesn't it), are private subsidiaries of Vigilant Solutions, a company that specializes in the sale of information to customers in law enforcement (Redmon, 2015). These companies appear to have fleets of vehicles that patrol cities capturing license plate information with remote readers. Their rationale appears to be that the car is in public, so privacy is not an issue. This is a tenuous assumption especially in light of the fact that the agencies appear to be reading tags parked in driveways, etc.

Despite what should be widespread knowledge that such activities occur, most people, as noted, simply click through without concern for what information is being collected, tracked and placed on computers. It is difficult to believe that these activities, or the by-product thereof, can go undetected by most members of the online community. Pop-up ads can be almost too on point, which only serves to demonstrate the effectiveness of data mining in revealing information about an individual.

To be fair, the vast majority of these tracking cookies and other information gathering methods are unlikely to endanger or imperil the user. The ads may instill a somewhat creepy feeling of being watched or monitored, which seems to be where most vocal complaints originate, but are not intended for nefarious purposes. Most of these are simply used for marketing, micro-targeting and the like.

This is not to say that the use of statistical models and data are new. More than fifty years ago, the UNIVAC (UNIVersal Automatic Computer) predicted the election of Eisenhower in 1956 (Roszak, 1986). This was a clumsy vacuum tube device that would be nothing by today's standards. Yet it was correct within 1 percent and called the election accurately based on a relatively small sample. Since that time, statistical modeling and the use of meta-data have progressed enormously.

The information that can be obtained through meta-data analysis is almost inconceivable. A recent study conducted at Cambridge University demonstrated that a wide array of information could be discerned about an individual through an analysis of Facebook "likes." Factors that were found related to "Likes" included intelligence and sexual orientation (Bean, 2013; Kosinski, Stillwell & Graepel, 2013: Kosinski et al., 2016). Another study by the National Academy of Science (2013) found similar results using Facebook "Likes" to determine sexual orientation, IQ and religious affiliation (Satter, 2013).

Unfortunately, the gathering of data does not stop with what had been voluntarily posted on social media. As previously noted, the information is simply too valuable to ignore. Facebook states in its terms of service that it tracks your location and also collects meta-data that others post about you or to you. Uber, the ride sharing service, not only tracks its customers it has been accused of threatening journalists with information gathered through its ride sharing service (Tufekci & King, 2014). Other researchers have taken a data mining approach to find a wealth of information (e.g., Moro, Rita & Valo, 2016) while other studies have determined personality factors based on the information (Azucar, Marengo & Settanni, 2018).

Google, for example, uses its massive power to comb every word of every Gmail and every Google doc, in order to more efficiently market its services to the consumer (Hansell, 2007). Few people are aware of this despite the fact that Microsoft ran a marketing campaign to draw attention to Google's practice. What exactly they do with the information is a proprietary secret. Corporate interests can hide behind this fence to obfuscate for what purposes the obtained information will be utilized. Trust us, it's all legitimate business, is the casual way such data collection is shrugged off.

Many individuals who think nothing of how their commercial data is utilized suddenly see the light when they discover that their data is used in the voting arena. One does not need be an anti-government extremist or a paranoid conspiracy freak to register a sense of unease that our voting patterns and preferences can be determined by such factors as what products we buy or what web searches we conduct. Yet that is exactly the case.

In a study concerning the buying habits and patterns of voters, substantial overlap was found between preferred products and voting choices. It was found that Starbucks customers were more likely to vote Democratic while Dunkin Donuts customers voted Republican. Choice of car brand, and even sports team preferences were found to relate to voting patterns (Lindstrom, 2008; Scott, 2015). Talk about voting with your wallet.

Instituting an opt-in approach would likely correct some of the issues with how individual companies collect data. But, as noted, this data is considered too valuable to ignore. The loophole, which the average consumer clearly does not understand, is that much of the data collected by them and used for various purposes is considered public behavior. Note that "public" has now been redefined to include what is done on a computer in the privacy of the home. In a now infamous example, Target, apparently determined a 16-year-old was pregnant due to computer searches and began marketing to her which eventually tipped off the father about her condition. While this case did not end in tragedy it is not a great leap to imagine scenarios where similar information may result in horrendous outcomes. However, what should be noted is that data collection never stops. Grocery store value cards, credit cards, phones, etc., all become data collection devices. The smart phone is the greatest collection device of all.

There are many other examples of information revealed through data mining. The technology and the math are so far ahead of the legal frameworks that data mining is a massive legal gray area. The security of personal information is not likely to change as the analysis of meta-data becomes more refined and the cumulative profiling of individuals reaches unprecedented levels. Already there are private companies and individuals, called data brokers, that have emerged to mine and sell meta-data for commercial purposes. These companies operate almost completely out of public view and with few laws to govern their behavior. This very lack of knowledge concerning the tracking of information is creating some alarm. According to Edith Ramirez, former chair of the Federal Communications Commission (FCC), "Consumers just simply don't know these practices are taking place" (PBS NewsHour, 2014, June 13).

The fact that the means by which algorithms and data mining works is poorly understood likely contributes to the lack of outcry over these methods. The unobtrusive and unnoticed manner by which this data is collected may also play a part in the acceptance of government surveillance. By and large these efforts are so seamless that most never notice them.

Not everyone, however, takes a laissez faire approach to data mining. Issues of location tracking, digital snooping, theft of personal data, and the assumptions that our online behavior constitutes public acts that allow corporate spying (tracking cookies), and other such info grabs in the name of commerce, are likely to have at least a minimal chilling effect on free speech and beyond.

For those who value privacy the digital age has created a level of uncertainty and insecurity that is unprecedented. In cyber world it is simply impossible to be completely secure. In the physical world one can usually buy enough locks or install a security system so that one can at least feel secure. To keep people from spying on you in the physical world the most extreme action most people need to take is to close the curtains.

It is at this point where the legal protections of the cyber and physical worlds diverge. In the physical world, if you lock your doors (and even if you don't), it is considered a crime to enter your residence. If you close your curtains, those who try to look beyond them are committing a crime. In the digital world this is not the case. The attitude seems to be the converse. There seems to be a prevailing ethos in the cyber world that assumes if you do not do enough to keep people out it is your own fault.

Even if one accepts the notion that the individual is responsible for providing their own protection against cybercrime, that does little to stop the onslaught of efforts to use digital technology as a means to commit these crimes. Most of the crimes committed using digital technology, through a plethora of means, phishing, Trojan horses, viruses, tracking cookies, etc., are geared towards financial gain. In these cases, and especially in cases involving bank data breaches and corporate and state espionage, it might be argued that there should be a reasonable expectation of efforts to steal the data and that there is a chance to prevent the thefts. However, efforts to fight digital crime are largely unsuccessful. Organizations with extensive operations to prevent theft are hacked on a regular basis. Nothing can really be viewed as safe.

Devices that can allow parents to check up on their children can be used by third parties to locate and track them. The Federal Trade Commission released a report in late 2012 that surveyed approximately 400 apps targeted towards use by children (Singer, 2012). More than half of these apps transmitted data about the children including their phone numbers, physical location and other identifying information. More thany three in five (68 percent) of these apps allowed for advertising to be targeted to the children. Very few of these "features" were disclosed in available information about the apps. This report did not fix the problem. In a more recent survey of devices, thousands of apps directed at children were found that violated the Online Privacy Act (Reyes et al., 2018).

In addition to these invasions of privacy issues, other identified hazards associated with children's use of apps include online fraud, discrimination and a number of safety issues including tracking by pedophiles (Singer, 2012, Flaherty, 2013). Every image file from a phone or camera has Exchangeable Image File Format (EXIF). These provide metadata that can include location where taken. The EXIF tags or geotags as they are sometimes called, are embedded in the photos and can provide information without any awareness on the part of the person posting. A simple posting of photos on the wrong site can lead to a pedophile being able to pinpoint the addresses of potential victims. Anything posted should be considered as existing forever. All it takes is a forward or a screen grab and there is no way to pull it back. That intended to be transient can become permanent. Must every parent be a digital security expert and cyber-hoverer in order to protect their children?

In the realm of data mining and data theft, or even in the realm of outright criminal behavior, it is easy to see how records and information can be compromised. What about those people that we have to trust for our digital security and operation of our computer systems? What about those who service the machines? Almost everyone who has ever had someone work on a computer experiences having to have the computer taken from the home by the technician. Even for those who do everything in the home, what assurances does anyone have that those who work on the systems do not have a sinister purpose in what they are doing. If you want to break into computers, let people give you access.

Medical records might also be susceptible to bad actors in industries such as the insurance business. What better way to get an advantage over competitors than to simply look at the files?

Medical Records

The move is on to digitize all records. Medical records were one of the first areas to prioritize the digitization of records. The idea was that patient care would be better coordinated and that the ability to coordinate and analyze large amount of data would lead to numerous medical breakthroughs.

Big Data does have the potential to provide mathematical models of disease processes, biomarkers, and treatment effectiveness (Saporito, 2013). However, the potential for this approach is just that—potential. Further complicating the picture is that the presence of biomarkers does not necessarily signal the presence of a disease, just the genetic potential. Rather than yielding perfect information and pointing to "cures" for diseases, statistical analysis is merely a means of determining probabilities. Genes are only one part of the equation. Environmental factors, iatrogens, teratogens, protein interactions, etc., also affect the final manifestation. Statistical analysis of the multiple factors that impact the expression of genes and diseases is at present a tentative process hampered by the limitations of statistical procedures.

One of the issues that is sure to emerge with the use of Big Data is that statistical probabilities become treated as fact. What may or may not happen in a predictive sense is only applicable to the group/population. For each individual the future is to some degree unknowable. A person may have the gene for breast cancer that makes it 85 percent sure that they will develop the disease, but if everyone with an 85 percent chance acts as if that is 100 percent and has a double mastectomy, then it will be impossible to isolate the protective factors. Ironically, acting on the information may serve to ultimately prevent finding a cure or even protective factors.

It is critical that health care providers and legislators get a handle on these changes in the manner in which health information is collected, maintained and used. In the relatively recent past, little beyond basic health information was or could be collected even by medical personnel. Now, it is possible for an individual to have their entire genome sequenced. According to the National Institutes of Health the average cost of sequencing an entire human genome plummeted by a magnitude that is almost inconceivable, falling from $10.5 *million* in 2006 to $6,600 in 2012 (Lee, 2013) to the levels found today where it is possible to get a DNA home test kit for less than $100 from a number of companies. The fact that the collection and analysis of such information is becoming cheaper by the day only adds to the likelihood of misuse. When it costs $10 million to sequence a genome there was little concern that this information could be widely abused. Insurance companies, for instance, might find it cost effective to demand such information.

The Surveillance State

In early 2013, reporter Glenn Greenwald and filmmaker Laura Poitras began receiving encrypted messages from a mysterious individual claiming to have earth shattering information concerning U.S. government surveillance programs. Face to face meetings were eventually arranged with the man, who was to become one of the most famous whistleblowers in history—Edward Snowden, a self-described "infrastructure analyst." Snowden was officially an employee of defense contractor Booz Allen Hamilton, but worked solely for the National Security Agency (NSA) according to his own report (Greenwald, 2014; Poitras, 2014). In essence, Snowden was a spy, a cyber spy, who came in from the digital void.

The information produced by Snowden went beyond the mere revealing of the contents of documents, emails and phone calls. It provided evidence of the existence of surveillance programs employed by the NSA and other government agencies that had either been denied or gone undetected. What Snowden revealed was not just information, it was the manner in which that information was being obtained and utilized. Snowden did not merely reveal the existence of espionage, he revealed the extent to which that espionage was directed at everyone and the lengths to which it is now possible to reach directly into the lives of every individual.

A number of programs were revealed that included activities such as wiretapping phones, monitoring email, data mining, meta data analysis that gathered email lists, phone lists and social media data under code names such as UDAQ, PRISM and Tempora. PRISM was a particularly pernicious effort in that it involved gathering data from companies including Google, Facebook, Apple, Skype, AOL, Yahoo!, YouTube and Microsoft, apparently without their knowledge and sought to utilize information from a number of sources including, but not limited to, emails, searches, blog posts, social network sites, etc. (Gelman & Soltani, 2013; Reuters 2013, October 30).

These revelations eventually led to the discovery that U.S. allies, including their heads of state, had been spied upon. Allegations suggested that as many as 35 world leaders, including Angela Merkel, whose cell phone was apparently tapped, and even the Pope, were under surveillance. The revelations came so fast and furious that it was difficult to keep up with them. In this respect, these allegations were similar to much of the digital world where information eventually becomes so plentiful that it dissolves into a mush with the essential issue—in this case the right to privacy and the freedom from unreasonable searches and seizures—becoming lost.

The details of these operations indicate that the collection of data was and is ubiquitous and recognizes no limits. The first of the revelations concerning the scope of these data collections efforts were reported in the *Guardian* newspaper when it obtained copies of a subpoena issued to Verizon that required the tracking of all phone calls in the United States. While the NSA stated that phone calls were not actually listened to, data about who called whom, the frequency and duration of the calls and other such information was collected (*Guardian*, June 6, 2013). The NSA largely ignored the greater concern that the ability to utilize meta-data collected through these methods is far more invasive. In effect, the NSA counted on the public to be too ignorant to understand how abusive these practices are.

Beyond the gathering of simple raw data, "meta-data," in effect data mining, of a wide array of digital communications has also been an ongoing reality since 9/11. The breadth and depth of the various data collection efforts and programs has been one of constant revelation. The scope of these activities is truly staggering. Inherent in the use of any technology is overreach. Rarely have humans developed any technology without trying to push it to its limits. The problem with data mining and other analyses of meta-data is that once a type of use is discovered it will likely be overused.

The ability to monitor individuals through a myriad of methods can now be taken to lengths heretofore unimagined. In an interview with NBC News, Snowden stated that the intelligence agencies of the United States (and China and Russia) can take control of a cell phone the moment it is turned on, using it to record, photograph and locate individuals. Snowden further claimed that the NSA has the capability to watch messages as they are being typed, keystroke by keystroke. This ability is more alarming than even the use of meta-data in that the ability to watch someone in the process of writing and editing is an insight into the person's thought processes.

Since the passing of the USA Patriot Act in 2001, there has been an ongoing outcry among civil libertarians and civil rights activists seeking to raise the alarm about the far flung consequences of FISA and the Foreign and Domestic Surveillance Act (FDSA), which allows, or more correctly, requires, the constant data mining of email, telephones, and other means of communication. These measures are so inclusive and so broad that it is difficult to see what could be excluded. If one pursues the idea of anyone connected to suspicious persons to its logical conclusion, as in the concept of "six degrees of separation" (discussed earlier), then it is clear that everyone is under watch.

In a report issued by Vodafone concerning government surveillance of its customers in nations around the world, it was determined that at least five nations—Albania, Egypt, Hungary, Ireland and Qatar—allow authorities "unfettered access." Several countries, such as the U.S., refused to allow Vodafone to have any information, citing disclosure laws. Britain and India appear to allow for the tapping of *all* phone companies in the two nations (Svensson, 2014b).

The information provided by Snowden was initially revealed slowly over time. Snowden's assertions were apparently confirmed with the WikiLeaks release of the CIA cyber "toolkit." These documents, called the "Vault 7 archive," exposed the vast arsenal of measures the CIA has for surveillance. Capabilities were revealed that include bypassing encryption and using common devices such as phones and televisions to monitor and record conversations. A vast array of cyber-attack programs, such as malware and tracking programs, were also uncovered (Shane, Mazzetti & Rosenberg, 2017). The revelation of these documents not only serves to confirm Snowden's assertions but also serves to demonstrate that there is almost nothing in the digital world that cannot be used to invade privacy. The fact that these abilities have been related to spy agencies provides no additional, or any less, paranoia: These capabilities may be possessed by various intelligence agencies but also by private individuals. In many ways, it is the private sector that should most alarm the average citizen.

Initially, Snowden chose to be an anonymous source out of a fear of retaliation by his bosses at the NSA. A film about Snowden, *Citizenfour* (Poitras, 2014), demonstrates his absolute expectation that his life was imperiled and that he expected the NSA to come for him. In a somewhat humorous and anxiety producing moment in the film, Snowden is discussing how phones can be monitored and realizes the hotel phone is plugged in. As he unplugs the phone a fire alarm sounds in the hotel. Everyone in the room laughs nervously.

Despite some changes to the Patriot Act as a result of an outcry, the exact nature of the changes are unclear and the Digital Security Act of 2015 *requires* the sharing of information by private companies with the government (Nakashima, 2015). Defenders of the program have pointed to its alleged effectiveness, which completely misses the point of the objections. The effectiveness of the program is irrelevant. A completely totalitarian state with no pretense of citizens' rights or process would undoubtedly be more effective at stopping all sorts of crime and misbehavior. It is exactly this sort of excuse—fascism works—that led to the creation of the Bill of Rights in the first place. The threat to constitutional liberties has been a contentious issue from the initial authorization of the Surveillance Act.

More alarming in many ways than those who defend the intrusion into rights on the basis of its effectiveness are those who defend widespread data collection on the basis of presumed innocence. This position seems to assume that just because you don't have anything to hide, it is okay to have everything you own, including your thoughts, searched.

The notion "if you're not doing anything wrong, you shouldn't mind being watched" is shocking at its most basic level. The very essence of a free society is to be free from the reach of the state. One does not have to be up to anything to want to be secure in their person, effects and thoughts. Must the whole world begin to act as if they live in East Germany circa the Cold War and that the Stasi is listening and watching everything that occurs.

Beyond both of these absurd arguments, (1) totalitarianism is effective and (2) only the guilty have anything to fear, there are other reasons to believe this "grab it all" approach is not grounded in liberty and reason. The fact that an IP address can be "spoofed," or manipulated in a manner as to mask the source, calls into question the legitimacy and integrity of all online transmissions and tracking activities. It becomes another avenue for questionable prosecution. Beyond guilt or innocence, there is simply more to the need for privacy and anonymity than a fear of persecution. Personal security and cyber safety are among the legitimate reasons for privacy. Transparency may be good for a society trying to ferret out terrorists, but it is life threatening for dissidents in repressive regimes.

The danger to the innocent, when spy programs "accidentally" sweep everyone up into their web, lies in the very means which allow security agencies to connect terrorists and criminals. The same patterns used to identify the dangerous can also be used to identify the courageous. In effect, the problems and the solutions in data mining and other forms of meta-data analysis are the same—everyone is connected.

When Senator Frank Church first used the term "surveillance state" in the 1970s he was referring to a type of government overreach that pales in comparison to what is happening today. At that time bugging phones and taping conversations was the limit of technological snooping. He could not have imagined the staggering levels to which this process has evolved. Certainly the information abuses revealed by the Church committee in the 1970s were invasive and a violation of privacy, but the use of meta-data, what Edward Snowden aptly called "pattern of life" data, can be even more revealing (Williams, 2014), allowing for the mapping of social networks among other uses.

The mining of data to determine a network of connections between phone numbers should be more troubling from a standpoint of civil liberties than actually listening to the calls themselves. Determining the connectivity between individuals is more invasive and telling than the actual conversations. The information gathered through the process of data mining that uses number crunching and various algorithms has the potential to reveal far more about a person than their conversations. It reveals their social networks. It reveals how a person lives and even thinks.

The Death of Privacy

Beyond the Keyboard: The Grid Is Everywhere

The death of privacy is unfortunately not confined to the activities of cyberspace. In the physical world it is becoming a near impossibility to be in a public (and sometimes private) space without being observed, tracked, and/or counted. From the use of radiofrequency (RF) tags, to the use of biometric data for identification on things like driver's licenses and passports, to the ubiquity of video cameras, life has become near constant observation. Alexa, the Amazon desktop AI, monitors and records conversations in the room in order to improve the experience of the customer. The scope of these intrusions into privacy has become so pervasive that it is almost irrational to assume that public behavior is not observed. The depth to which behavior is recorded and stored reaches levels that were impossible only a few years ago. Some of this data collection is surreptitious while other forms are so ubiquitous they are fading from conscious awareness.

Video Cameras

One of the more obvious and visible means by which privacy is rendered a quaint notion of the past is through the increasing presence of video cameras in public places. Where there are not cameras, there are always cell phones. Cameras have become so commonplace that they are nearly expected. From gas stations to banks, from toll roads to parking lots, from big box stores guarding against shoplifting to parks guarding against muggings, cameras, both public cameras placed by municipalities, law enforcement and other governmental groups, and private cameras placed by corporate and commercial interests, have come to be expected. The NYPD has pioneered the use of smart cameras for monitoring public spaces. Algorithms are employed to determine such matters as whether a person has been sitting for longer than three minutes in the same place. In effect, the cameras decide if you are acting suspicious. As common as cameras have become in America, the U.S. barely holds a candle to those in other places.

China is reported to have the most observed populace in the world with over 200,000,000 cameras (PBS, 2019 Sept 30) which are equipped with the application of artificial intelligence programs used to gather information, engage in facial recognition and develop social profiles of people from actions in public spaces. These programs can be accessed by government officials and are being exported to other nations such as the Philippines (PBS, 2019 Oct 1).

London is likely one of the most watched cities in the world outside of China. Estimates vary substantially as to the number of cameras, but it is difficult to walk anywhere in the city without being recorded. A case in point, the London Subway bombing culprits were rapidly apprehended primarily because they were tracked from the point of the bombings to their homes with near seamless recording. The poisoning case involving Russian spies was assisted by tracking the culprits almost every moment they were in the country. Similarly, in the bombing at the 2013 Boston Marathon, cameras and other electronic information such as cell phone pings, were instrumental in providing the evidence that led to the apprehension of the culprits.

During a riot in Vancouver following a hockey game, police obtained more than 5000 hours of video, identified over 15,000 crimes and arrested approximately 350 people (PBS NewsHour 2013, April 17). In the past, only a few of these people would likely have been apprehended. There is some evidence emerging as to how behavior is altered by the presence of a camera. People tend to go to extremes and either ignore or perform for the camera. There is little room for middle ground, it seems, in a digital world driven by a forced choice of "0" or "1."

It is likely that sports events like the Super Bowl or celebrations such as Mardi Gras are provoking greater displays of excess. On the other hand, traditions such as taking a piece of turf after the World Series are disappearing because fans realistically fear they will be caught and persecuted for such behavior.

Public awareness that video has led to crimes being solved has no doubt led to what appears to be overwhelming public support for the presence of cameras with very little apparent concern over civil liberties (Schuck, 2017; Weisman, 2013). Support for cameras, especially in high crime areas, is strong. Of course, it is in the area of civil liberties that cameras are prone to provide overreach. For a sense of how this can be willfully abused one should review Xu, Hu & Mei (2016), which practically gushes over the possibility of providing complete surveillance. A further review of the various patents for cameras, which record far more than images, many employing various types of artificial intelligence, can be seen at patents. google.com/patent/USD74087151/en.

Video cameras present a number of challenges not only to privacy but to the very ability to live a life free of interference. To further illustrate the point that video surveillance can be used for oppressive intent, one need only look at the case of China and its increasing reliance upon such tactics. Chinese state media recently announced that "every corner" of Beijing is covered by surveillance cameras (Yin, 2015). The purposes of these cameras appear almost solely for the intent of altering or chilling public behavior and to keep tabs on dissidents.

It is not just state and public surveillance that the digital age has brought upon us. The case of Dharun Ravi brings to the discussion the notion that even private behavior can be observed. Ravi was a college student who remotely activated a camera in his dorm room in an effort to watch his roommate engage in sexual behavior. Ravi's actions likely were motivated more by immaturity than by malicious intent but led to his prosecution when his roommate committed suicide (Cook, 2012).

Remote activation of cameras, which many people do not even know is possible, is a relatively simple process. In fact, the more connected the world and its various digital devices become, the greater the chances that these devices can be compromised. The Internet of Things, whereby all manner of devices are connected for remote control and access, is an open avenue for invasions of privacy. Security flaws in numerous devices have been exploited, allowing hackers access to private cameras used to monitor babies and security systems (Wyatt, 2013). Other hackers have tapped into systems via utility company monitoring equipment along with remotes to open garage doors and other means of physical entry (EurActiv.com, 2014). Security researchers at Georgia Tech were able to hack iPhones and iPads using nothing more than USB chargers (Greenberg, 2013; Szczys, 2013). Note that all of these events happened several years ago. The ability to breach security measures has only become more advanced.

Drones: The Eye in the Sky

Drones, until recently, were mostly known for their military applications but are the wave of future invasions of privacy. With a camera attached to a small mobile flying device that can be remote controlled the possibilities are seemingly endless. Heralded as a means to deliver packages, map remote areas, engage in research, and other laudable goals, the true value for many is the ability to attach a camera to the device. The legitimate and valuable uses of camera equipped drones are numerous and include border patrol, farm research, biological research, fighting animal poaching, monitoring hotspots in fires and volcanoes, and mapping activities among others. However, despite all of these uses it is almost assured that the technology will reach its height with paparazzi spying on celebrities, real-estate developers finding the last undisturbed land and 13-year-old boys peeping on their neighbors. Corporate spying is almost assuredly happening. The fact that the price structure is relatively cheap, and the technology is fairly easily mastered only increases the potential for abuse. It is estimated there may be as many as seven million drones in use in the U.S. by 2020 (Gonzalez, 2016). Couple this with the fact that there is little regulation of these devices and there is a recipe for disaster with drones and cameras as the main ingredients. The few regulations that exist are more centered on registration than on the actual operation. Issues of use for surveillance by law enforcement have barely been raised. More significant concerns such as use for malicious and malevolent purposes are poorly addressed. An example of the very real problems produced by what is essentially a toy can be found in reports from the Middle East. In the battle for Mosul, Da'esh (Islamic State) reportedly used drones that they had bought "off-the-shelf," to drop bombs on U.S. and Iraqi forces (*Face the Nation*, 2017, February 26). The military applications of drones are sure to be a major threat in the near future. It seems improbable that such an inexpensive and potentially lethal device will not be used by every military on the planet.

Databases, Public Utilities and Other "Helpful" Features of the Digital Age

A source of information commonly manipulated for ill intent are those records that exist as a public service. The amount of information contained in these can be enormous. Some of the sources are not only available to the public, their entire function is to provide information. Some databases, such as sex offender registries, have a clear purpose. Others, particularly those related to taxes and real estate matters, are available as a result of public records laws. Many of these are utilized by the private sector in a manner not exactly as intended by public disclosure laws. It is the collation of various sources of information that routinely serves to violate privacy.

Real estate sites provide a clear example of this. While much of the information available on this site is valuable to a potential homeowner it also allows anyone to determine a great deal of information about nearby houses and properties. With little effort it is possible to determine matters such as the approximate value of any home, the amount of taxes paid, sales history of the property, school district, and many other bits of information. This is true even for homes that are not for sale.

It is nearly impossible to enter an address into a search engine and not be able to see the actual house, at least an aerial view if not a street view as well, and a survey of the lot. There may be no actual "harm" from this unless privacy is a concern. This data clearly has commercial value, yet most homeowners are unaware such information is available. Almost no one appears to expect compensation.

Beyond real estate matters and sex offender registries there are enormous databases waiting to be tapped and mined for information. Everything is seemingly available and with a little effort and savvy there is little that is out of reach. The potential for abuse seems obvious. Police departments in many municipalities utilize a program called BEWARE to crawl information in public databases, commercial information and even social media to develop a "threat rating" for individuals. This information can be assessed in real time for purposes such as traffic stops (Skorup, 2014). The legality of this snooping seems to be unchallenged, largely due to a lack of awareness that it even exists. Possibly having an even greater potential for abuse is that information which is not exactly public information but is rather gathered in the course of public activity.

Collected Information:
The Sensor Grid Is Watching

In addition to databases, data mining, overt recording by cameras, and cyber surveillance, there is also the issue of clandestinely collected information related to public action. One of the more insidious means by which privacy is undermined relates to the fact that technology has progressed to the point that collection and storage has become relatively easy. Far more data is collected and stored for the simple reason that it is possible to do so. At present, without a great deal of public awareness, vast sensor networks have been installed in a multitude of places. These networks collect data of any conceivable nature. The reader is encouraged to search for online demonstrations of how sensor grids work. Many of these networks can be publicly accessed and the information can be observed in real time.

These sensors networks can monitor any manner of parameters from motion to temperature. Much of this collection appears almost pointless at present as researchers work to develop the technology to process and interpret the data (Dublon & Paradiso, 2014). This lack of a defined use of the data leads to enormous amounts of information being stored or "siloed" waiting for a future use. The fact that this can be done without the knowledge of those involved should be alarming.

The expectation that one's behavior is being recorded is increasing. Even when unsure as to whether anyone is watching, the modern citizen should likely act as if their behavior is being monitored. It is becoming almost impossible not to be monitored.

On occasion the sensor grid is employed for purposes that seem legitimate but are still done surreptitiously. During the Ebola outbreak of 2014–2015 China began using body temperature sensors to screen people at airports. Footage of this process appears to show masses of people being scanned without their knowledge or awareness (PBS NewsHour, 2014, July 31).

Even the Dictators Are Watched

No one is immune to the constant and ubiquitous surveillance that pervades the existence of every human on earth. Consider the case of Guatemalan strongman Jose Efrain Rios Montt. In April 2013 he became the first former head of state tried for genocide in a domestic court (Marder, 2013). Accused of ordering genocide against the Mayan population, much of the evidence against him is the product of forensic science and surveillance technology. Some of the evidence that supports the case for genocide is geographic data recorded by satellites showing the movement and disappearance of people and their settlements.

In a similar vein, cell phone cameras and other data provide evidence of criminal wrongdoing on the part of those in power in a way that could never have been done in even the very recent past. During Watergate, the American public was surprised to learn that Nixon had taped his own office. Back in the 1970s tape recording was the height of surveillance. Now it is difficult for anyone, even dictators, to act in secret.

Of course, few dictators fear retribution. Most are semi-deluded enough to believe that their actions are justified or necessary. While they may be watched and more of their actions documented, in the long run, those who control the media and the surveillance technology are more likely to use it against their populace than to have it used against them. One of the reasons Rios Montt is the first dictator to be tried in a domestic court for genocide is that most dictators go out in a blaze or they go out in an airplane, fleeing à la Idi Amin, to a place (and palace) promising not to send them back.

The most egregious of dictators, it must be realized actively seek to control such means for their own purposes. Even though a random person in a crowd in Syria can document atrocities with a cell phone, it is still more likely that those who control the technology will be the ones who use it to their benefit. The signal that transmits the message is the same signal used to track the dissident. The arms race never stops; from sticks to stone, to atom bombs, to cell phones and satellite signals, the race for weapons technology is endless. No one is immune.

The capability of satellites has reached the point that license plates can be read from space. Not only are those involved in the technological age affected, so is everyone. "Stone Age" tribes, with no contact with the modern world, can be watched as well. The fact that they don't know they are being observed in no way mitigates the action. It may well be that this makes it worse. Those living in the technologically advanced parts of the world likely should no longer have any expectation of privacy. Not having awareness of such technology would seem to make the violation all the more pernicious. To those for whom this point is not self-evident, perhaps there awaits a career in politics.

The technological age is quickly outpacing the ability to consider its implications. This is nowhere more apparent than in the area of privacy and private information. Not only can privacy be invaded without the consent or knowledge of an individual, it is now possible to take entirely physical information and determine factors about a person against their will. In years gone by the Supreme Court once ruled that there is a "reasonable expectation of privacy" for all people. Today that seems a quaint view of an antiquated era. Privacy is dead.

The Pope Is Also Under Surveillance

The papal conclave following the resignation of Pope Benedict XVI illustrates the perils of life in a time of ever developing technology. Prior to the beginning of the conclave, security personnel swept St. Peter's Basilica and the Sistine Chapel for devices that could be used to eavesdrop, record or otherwise spy on the proceedings. When the College of Cardinals becomes as paranoid as Mafia bosses about their privacy, there can be no pretense that anyone can be secure in their person.

Several recent court cases have put checks on technology, preventing the police from using thermal imaging (Kyollo v. United States, 533 U.S. 27) and GPS tracking (United States v. Jones, 132 SCOTUS 945, 565 U.S. 2012, No. 10–1259) and other such approaches to search willy-nilly for illegal activity, but allowing it for data-mining of health information in a case from Vermont (Sorrell v. IMS Health, 131 SCOTUS 2653 U.S. 2011, No. 10–779). However, much of the technology formerly associated only with law enforcement and espionage has now become affordable and available to the masses.

Neighbors can now, right now, spy on each other. Corporations can easily gain access to the buying patterns of consumers. Political campaigns can find every voter who might be persuadable and micro-target them. All that is required to do these things it seems is the will. And if someone has the will, there is nothing that can be done about it. Again, even the pope is not immune. It was reported that a drone carrying a high-resolution camera was discovered over the Vatican shortly after Christmas in 2015. Two individuals were taken into custody in an event that received scant attention (Hartman, 2015).

Worrying about your neighbors spying on you is probably not your biggest concern. Their interest is likely prurient more than malicious. Those who are engaged in nefarious and otherwise criminal activity are of greater concern. Organized crime groups, both in the United States and internationally, have discovered there is money to be made from cybercrime. All other reasons aside, be it from concerns of government surveillance or the presence of tracking cookies, it is those who have malicious intent that are the greatest cyber threat. No matter what one fears, all efforts to protect digital material are ultimately futile. Security in cyberspace is an illusion.

The Great Legal Gray Area

Who owns your data? The naïve answer is to say, "I do." Legally that may not be so certain. But why not? How is it that this is not theft at some level? How is it that it seems legal to get in your computer and look around? How is that not breaking and entering?

Jaron Lanier argues, in his book *Who Owns the Future?* (2014), that if a person's commercial data is considered so valuable then the individual who generates the data, not the one who breaks into the computer and takes it, should be compensated. The basic idea is that it is your data, your commercially valuable data, and there is no other circumstance where the person who produces that product is not compensated except under conditions of slavery and theft. As things now stand, information is the property of the person who collects it or steals it. The thief appears to be rewarded with compensation.

According to Lanier, the alternate view, that everyone who contributes to the digital network should be compensated originated years ago with Ted Nelson but has not taken hold—primarily because of the hidden nature of data collection (PBS News Hour, June 17, 2013). Charging for the collection of data by tracking cookies or the taking of images on public cameras is parallel to the manner in which all other content is protected and compensated. The act of charging for data that is now mined would likely substantially change the manner of data collection in a far greater way than any legal means. At present, logic does not prevail.

Who owns your data or for that matter your biological information is not clear-cut in the legal sense. Technology has gotten so far out in front of our ability to think about it that there is considerable lag in the legal arena as regards rights, boundaries and privileges.

Funding to examine all the resulting ethical, legal and social questions that might result as a consequence of unraveling our genetic code was paltry, with only 5 percent of the total budget going to look at these issues (National Human Genome Research Institute, 2003). Because of an essential lack of funding for these matters, and an accelerated scientific process, a situation is produced leading to an ability to do something that may be fundamentally wrong, without any process to even examine such acts.

Several questions are rightfully raised by the science:

> Who owns each person's biological code?
> What can each person do with their own code?
> Can others have access to the code?
> Can it be patented, copyrighted, etc.?
> Who will be allowed to manipulate genetic material?
> Can a person be ordered to alter their own code?
> Can your genetic code be used as evidence against you?

These questions only begin to scratch the surface of the ethical issues raised by the ability to decode and in some cases manipulate DNA. In retrospect, it might have been better if 5 percent of the money had been spent on the science and the bulk on the social, ethical, and legal ramifications. Like the bomb, this genie is out of the bottle as well.

Significant issues have already arisen regarding these concerns. One of the more celebrated cases involves that of Henrietta Lacks. Lacks died from a particularly virulent cancer and had tissue samples taken without her knowledge. Cultured in a lab, the cells grew, well, like cancer, and produced a tremendous amount of biological material.

Who Owns Your Biological Data?

The case of Henrietta Lacks, first brought to public attention by Rebecca Skloot in her book *The Immortal Life of Henrietta Lacks* (2010), provides a stark example of the means by which technology can not only invade privacy but perpetuate a wrong. Ms. Lacks, who died of cervical cancer in 1951, was the source of a set of cancer cells that grew vigorously in a laboratory environment. These cells became known as the HeLa line and were used by researchers around the world to conduct research on cancer cells. All of this occurred despite a lack of consent or even knowledge by Lacks that these cells were collected. Despite what became an enormous contribution to medical science, neither Henrietta Lacks nor her descendants were ever compensated or made aware of their use. Discovery by family members of the cells' existence was largely a case of serendipity (Skloot, 2010).

Lacks is not the only individual from whom such genetic information has been gathered or extracted under lack of consent or even deception. The Havasupai tribe in Arizona was for years involved in a contentious lawsuit with Arizona State University over genetic sampling and a consent procedure largely viewed as either incomplete or deceptive. The details of the case are not in dispute. Arizona State researchers, in an effort to study diabetes and markers for the disease, began a study using members of the Havasupai tribe. Diabetes is rampant among the Havasupai and the tribe agreed to be studied in hopes of improving the health of its members. Information taken from the tribal members was then not only used for study of traits and genes related to diabetes but also analyzed for other genetic information (Henaghan, 2014). This case would almost assuredly not been discovered except that a member of the Havasupai tribe was enrolled in a class at the university where a professor began to explain the results of the testing. This case eventually snowballed to the point that employees of Arizona State were banned from all tribal lands (Henaghan, 2014). Long after the issue first became a serious conflict, researchers from both Arizona State and the University of Arizona continued to publish studies using this information.

The use of DNA to identify individuals is becoming easier and cheaper, leading to more frequent collection of the information. Federal prisoners and many state prisoners have their DNA sampled as a matter of course. These samples are then compared to databases to find matches to DNA left at the sites of unsolved crimes. Numerous individuals have been apprehended for crimes committed years before. In many of these cases the individuals were not even suspects.

The use of DNA cuts both ways as pertains to criminal culpability. The Innocence Project, a group that seeks to free those wrongfully convicted of crimes, has used DNA evidence as exculpatory in at least 365 crimes as of May 2019 and helped identify 160 suspects (Innocence Project, 2019). In many of these cases the individuals have served many years in prison for crimes that were committed prior to the ability to analyze DNA. In many of these cases there has been resistance on the part of the courts to re-examine the evidence.

Despite the reluctance to analyze evidence for exculpatory reasons, the collection of DNA for other purposes is becoming more routine and codified. In Maryland v. King, the Supreme Court of the United States upheld the collection of DNA samples from those who had arrested but not convicted of crimes (Aronson, 2013). Nearly 30 states had already been collecting this data despite legal objections. Justice Antonin Scalia, in a very strongly worded dissent that was joined by three of the liberal justices, expressed concern that the ruling could lead to the creation of massive DNA databases on all of society (Maryland v. King, SCOTUS).

More than 50 nations already have some form of national DNA database for police purposes. The United States has the largest with the FBI's Combined DNA Index System (CODIS) holding the DNA of more than 11 million people. The United Arab Emirates is developing a database holding the DNA of everyone in that nation (Lawless, 2013). Likely the greatest coverage is in several Scandinavian countries which have large databases on most of the nation.

The information that can be gathered through DNA can be monumental. Individuals and their families have been identified through supposedly anonymous data that was gathered for research (Gymrek, McGuire, Golan, Halperin & Erlich, 2013). The "Golden State" serial killer was discovered through the use of DNA matches to relatives who used genealogy sites (Regalado, 2018). The use of genetic material can be used for matters far beyond the use of solving crimes. It is a very small leap from collection of DNA samples for the purposes of solving crimes to the use of DNA by insurance companies to determine premiums. Given that insurance companies are private entities that routinely request very sensitive information, it is difficult to see how this practice could be prohibited in the U.S. except through Congressional legislation. What will be the legal guidelines surrounding this issue? Will they favor the individual or the corporate entity? As with the case of meta-data, those with the money and lobbyists will have a far greater probability of getting their voices heard than the average citizen.

In addition to the above, genealogical testing, now that it has become affordable, has also become more widely used. There have been numerous unexpected problems arising from this widespread use including such matters as discovering that one is adopted, that one's biological parent is different from that believed, that there are unknown family members in the world, etc.

Being monitored has become a part of life in the digital age. Big Brother now has many forms, be it social media, Big Data, tracking cookies, foreign and domestic surveillance programs, corporate data mining, security cameras, cell phone triangulation and a myriad of other methods. It is a paranoid's nightmare realized, and an aspiring celebrity's dream world. Whether a person chooses to live online and record every aspect of their life or move to a remote island and attempt to check out, the fact that we are monitored, and know it, is altering how we live. The machines are now changing us. They may be changing us so much they are affecting our very existence. One thing is certain, whether your fear of a loss of privacy is due to concerns about governmental overreach, fears of corporate snooping, meta-data mining, or just a nosy neighbor, cyber-spying is a reality. It is not just that there is a perception of a lack of privacy and security, it is a factual problem. There has already been at least one court case involving the gathering of DNA by a private company to investigate its own employees (Visser, 2015).

Whether at the level of nation-state or a corporation or an individual who is nosy and checking up on another person, there *is* a way to do it. All it really takes to spy on a person in the digital age is the will (and a little knowledge).

Should governments magically cease the gathering of information, the vast enterprise of data mining will continue as the corporate world begins to see the monetary value of such data and the multiple commercial applications. Terms of Service Agreements will continue to be filled with legalese and verbiage that ensures few really understand what is being given away. Unfortunately, cyber spying and DNA sampling are not the end of the ways that technology can be used to undermine privacy. Many methods of surveillance and data collection have nothing to do with being at a computer keyboard or having medical procedures conducted. Much of the data that is collected and stored is simply a fact of life. Being recorded and cataloged is life in the 21st century. The grid is everywhere.

The awareness of being tracked must necessarily have some effect on human behavior. Given what has transpired, it is difficult to believe that an informed person could ever have any expectation that what they do online is private and secure.

The expectation of privacy is dead and buried.

9

Artificial Intelligence

Meet Your Android Overlords

The journey of 10,000 miles begins with the first step.—Lao Tzu

At a very basic level artificial intelligence is already a reality. Based on a definition of AI that is grounded in the ability of machines to analyze information, make decisions, modify actions based on experience (i.e., learn), interact with other intelligent beings and operate independently, we are there. From a purely technical point of view "thinking machines" already exist. An example of technology that would meet these criteria is that which is commonly referred to as a voice assistant. Not only can the smart device and the associated voice carry out tasks such as interpreting meaning and modifying actions based on experience, but many users interact with it as if it is a living being. Despite many of these devices' having names, they are still just machines. For most people, artificial intelligence is more than just a machine that carries out complex functions. Eventually it might be, but at present, the machines are still just hunks of metal, plastic and toxic chemicals.

Machines that meet technical definitions are not what most individuals view as AI. The Holy Grail, or the alchemist's gold, in artificial intelligence is the attainment of a sentient, thinking, conscious, and self-aware being. This idea may well be the final frontier of AI, and without a doubt, there are people currently working to make this a reality. At present, however, this journey is still in the formative steps.

Before we are able to create an android such as Mr. Data from *Star Trek*, or Roy, the Nexus 6 from *Blade Runner*, there will be many intermediate stages that raise both technical and ethical challenges. From a certain point of view, the realization of this type of AI is much further away than is presumed. Paradoxically, from another viewpoint it is closer than we might think.

The broad range of what is meant by AI can be confusing in and of itself. In discussions outside the scientific literature, there is often a failure to define terms and as a result there may be no differentiation made between the use of AI as a means to control a thermostat and AI as a human replicant. Clearly, such a distinction is important.

The origins of AI might be traced to Joseph Marie Jacquard, with the invention of the programmable loom in 1804. The ability for a machine to make what had previously required a person ushered in the age of automation. At a basic level what Jacquard created was a robot that could weave fabric. The displacement of workers also began with this step. The development of the train and the automobile, while not usually thought of as robots, might be conceptualized as mechanized equivalents of the horse. Mechanization is a necessary precursor to artificial intelligence in that organic beings are removed from the process. The assembly line adds another dimension of automation. While seemingly disconnected from current conceptualizations of AI, these efforts were the first steps on the journey. Robots might be viewed as the next step.

Robots

The term "robot" has undergone significant reinterpretation and redefinition over the past hundred or so years. The origin is a Slavonic word, *rabota*, and means servitude or forced labor and is found in the play *RUR* (1920) by Czech playwright Karel Čapek (Flatow, 2011). Ironically, or perhaps prophetically, in the play the "robots" were mass produced organic beings who lacked souls and emotions.

The conceptualization of robots in popular culture quickly shifted to that of machine. By 1927, in Fritz Lang's *Metropolis*, the use of mechanical robots for sex, or sexbots, makes its initial appearance. From that period forward the term robot seemed to apply to mechanical beings who performed human functions. Robots at this point were more equivalent to what are now called androids.

The development of the field of robotics, in many ways ushered in by the space race, changed the use of the word. Robotics introduced an element of control at a distance. The development of machine learning—i.e., the ability of machines to alter programming based on experience—has further altered the concept. At present, "robot" has connotations of an unadaptable, basic machine that carries out repetitive functions. Applications for robots have been for such disparate purposes as assembly line work to surgery. Robots have even been developed to perform such mundane tasks as installing drywall on construction sites.

Uses of robots for mining, bomb disposal, and space exploration often involve machines that bear little resemblance to a human being. While the resemblance to a human being is irrelevant to the levels of AI a machine might possess, humans nonetheless process reactions towards AI largely based on appearance. The AI used to guide airplanes, perform elements of surgery, control pop-up ads, etc., may contain far more computing power and more qualities related to a type of intelligence than a robot that can put up drywall, but the latter seems far more disconcerting that the former primarily due to its similarity to a person.

Some robots clearly are being designed for "demonstration purposes." Activities such as playing trumpets, jumping, etc., are clearly useless activities for robots. The primary purpose of these developments psychologically seems to be to humanize machines. This effort is more related to the acceptance of them in physical space than any assessment of capabilities.

The real question is not whether a robot looks like a human or whether it looks like a vacuum cleaner. Theoretically, the idea of a transformer, as in car to fighting machine, matches perfectly the idea of a fully realized artificial intelligence. However, getting from robot to transformer is not just a matter of computing power.

Moving from machines to artificial intelligence requires more than a step, it requires a leap. Before a machine is viewed as intelligent, it must be capable of learning. The ability to learn is far more than an aggregation of knowledge. Learning requires the ability to modify knowledge based on experience. Humans and other animals possess this ability. For artificial intelligence, this is the essential hurdle—learning.

Machine Learning

The term "machine learning," is something of a misnomer if one is imbuing the device with the ability to think in a manner similar to the human brain. However, if one is describing learning as the ability to modify outputs based on previous inputs or experiences then the term is applicable. Of course, the term also has utility in that it rebrands AI, making it more palatable.

At the core of all machine learning are algorithms. Algorithms are the rules by which computers sort information. The machine then modifies its program based on the application of patterns and information discovered by the algorithm. The larger the data sets employed, the more accurate the "learning." At base, what all algorithms attempt to do is categorize information. Through this method machines can be instructed to look for certain patterns (supervised learning) or to use a trial and error approach that seeks to discover patterns (unsupervised learning). These methods are used in search engines, music services, speech recognition, and numerous other applications. Pop-up ads, tailored to the individual, are a function of algorithms.

Machine learning, at its most simple, is nothing more than the application of the patterns uncovered by the algorithm that further modifies the algorithm. Experience—i.e., more data—tends to increase learning. Devices such as voice assistants, initially experience a strong learning curve and become more "accurate." Systems such as YouTube and Spotify also use this approach to home in on what the consumer wants, to "enhance" the experience. However, beyond a certain level the approach starts to reach a point of diminishing returns and the results can become predictable and boring.

The use of the word "consumer" draws attention to one of the more trenchant realities of machine learning and what will certainly be the case for AI as it reaches greater levels of complexity. The commercialization and commodification of the technology is already obvious and will only continue to increase. This technology will almost assuredly rest in the hands of large corporations and those with wealth and power will control and exploit it. An example of this is apparent in efforts of Larry Page of Google to corner the flying car market (Harris, 2018). While the masses may have "access" to algorithms for use, it is for the commercial exploitation of and use of personal data, that "access" is granted.

The way the machine "learns" is by taking the data of users and constantly resorting and recategorizing it. This is the underlying method by which all algorithms work. It is pattern recognition applied over and over, ad infinitum. For purposes such as marketing, this process may be viewed as inconsequential or even helpful by some. However, this same process can have tragic consequences when incorrectly applied.

Abuses of this technology have already surfaced. Facial recognition programs, touted as existing for security purposes, have already been shown to be biased against those with darker skin (Buolamwini & Gebru, 2018) while predictive policing programs have also demonstrated severe racial biases (Lum & Isaac, 2016).

It is just such problems that point to the dangers inherent in turning over decisions to machines. Here again, the human worship of technology leads to incorrect assumptions about the ability of AI to make correct decisions. Numerous consequences and abuses of the inappropriate use of a machine learning algorithm are easily envisioned. Rejection for credit, misidentification in background checks for employment, incorrect medical diagnoses, and so on are all within the realm of current possibilities.

Unfortunately, an algorithm is not perfect. It is nothing more than statistical probabilities. Data mining can find connections even when there are none. Correlation does *not* equal causation. Yet, data mining assumes that it does. It has been noted that data mining has found relationships between Google searches and criminal activity, stock prices, heart attacks, election results, and even random numbers (Smith, 2019).

Despite these abuses and by extension the problems inherent in the application of increasingly sophisticated machines, at both the technical and ethical levels, development is full steam (or kilowatt) ahead.

Deep Learning

One step beyond machine learning is the process referred to as "deep learning." This is often explained as an effort to duplicate the neural net of the human brain. This analogy is simply inappropriate. The comparison of the human brain to a computer is strained at best. It may fit the techno zeitgeist, but it serves only to create false comparisons. Computers are viewed by some as smarter because the knowledge base is greater. A library has a greater knowledge base as well, but we do not usually consider buildings "smart." Likewise, a computer may do calculations much faster than a human, but that does not give understanding to the result.

What deep learning attempts to mimic in terms of the way the human brain operates is the ability to engage in parallel processing across multiple layers of inputs. Computers are great at serial processing, one thing at a time, but are not so good at coordinating multiple inputs across different levels of processing. It is in this area that the human brain excels. Developers of AI are attempting to use this framework of the neural net to further deep learning. The essential task is to get the machine not only to learn based on experience but to use other processes to further modify the algorithm.

The recent history of "thinking machines" is one of rapid advance. It was only in 2011 that Watson, an IBM computer essentially programmed with the cumulative encyclopedic knowledge of human history, beat Ken Jennings the all-time *Jeopardy* champion. DeepBlue, a computer that was programmed to play chess, using an algorithmic approach that sorted through all the possibilities of moves and countermoves, overwhelmed world champion Gary Kasparov through a method of mathematical exhaustion (PBS News, 2016, January 27). However, both Watson and DeepBlue were essentially without the capacity to make judgments and worked only as coded.

In what can only be heralded as a major breakthrough in artificial intelligence the Google DeepMind project developed a deep learning approach that appears to be able not only to alter its program based on past experiences, but also to engage in intuitive processes. Using the game of Go as the teaching paradigm, the computer was eventually able to beat the world's best human players (Silver et al., 2016). Go, unlike chess, has almost unlimited possibilities of moves and countermoves. Played on a 19 × 19 grid pattern upon which are placed small black and white discs ("stones"), players attempt to control areas of the board by boxing-in spaces. The game can change rapidly, and even experienced players have difficulty knowing exactly what moves to make. In short, the game requires some degree of intuition to play.

Humans apparently still have the advantage when it comes to poker. It seems that bluffing is the one skill the machines have yet to be able to decipher. Efforts continue on this and other fronts. The most hyped version of AI at present is autonomous vehicles.

Autonomous Vehicles

Numerous corporate entities from major car companies to Google are working on the development of autonomous vehicles. Generally thought of as "driverless cars," autonomous vehicles are much more. Already in use or soon to be are buses (SF Bay Area), cabs (Shanghai), delivery drones (Amazon, Windfield & Scott, 2016), bullet trains, and delivery trucks.

The use of AI features in airplanes has for some time been far beyond that used in any of the applications above. The term "autopilot" clearly defines the essential feature of autonomous technology. Unfortunately, aviation has also provided the most cogent examples of the perils of letting the technology take over. The crashes of the Boeing 737's in the recent past have mostly been attributed to the autopilot overriding the human operators (FAA, 2019).

To work, autonomous vehicles must function perfectly. Consider all the difficulties regular cars have with recalls and mechanical failures and suddenly the widespread use of autonomous vehicles can seem unrealistic. Amazon has put forth a plan to develop flying, autonomous warehouses (Fung, 2016). The idea seems fanciful if not outright ridiculous. While the effort to increase profits seems to suggest that anything is viable, flying warehouses are about as likely as world peace. The idea of a warehouse crashing out of the sky aside, there are significant concerns as regards the potential for accidents.

Driverless cars have demonstrated some of the same problems as those in aviation. In early tests, driverless cars experienced 272 failures and 13 near misses that required human intervention (*The Guardian*, 2016). In a test in California, a Google car hit a public bus (Newcomb, 2016). In

Arizona, a pedestrian was killed by an autonomous Uber (Wakabayashi, 2018). The promise of these vehicles has followed the usual pattern of hype and overpromise. In a similar vein, the UAE announced plans to develop a fleet of autonomous aerial vehicles, flying taxis, that would be in operation by summer of 2017 (Goldman, 2017). After heralding the first successful test in September 2017 there were predictions that the vehicles would be operational by 2018 (Murphy, 2018a). Less than four months later the operational date had already been pushed to 2020.

For technologists, these issues are just problems to be solved. For lawyers around the world, this is an opportunity. What has already been realized is that since there is no "driver" to be held responsible, the legal responsibility will default to the manufacturer. Even with matters as simple as mechanical malfunction this could prove a legal quagmire. Beyond the issue of malfunctions, there are also concerns as to what decisions the AI in control might have to make. Autonomous technology will necessarily require decisions with ethical implications. For example, suppose the choice is to illegally pass a school bus or to hit a person. What decision does the AI make? If the choice is hit a bus loaded with children or run over an animal, what should the car do? What about the choice between hitting an elderly pedestrian versus a mother with a stroller? While all these choices could confront a person, the AI will be rule bound and limited by its program. In the case of the school bus full of children or an animal, if the program prioritizes hitting objects over living things, it would hit the school bus to avoid a snake in the road. The essence of the problem is that an AI is only able to follow its programming. Novel and ambiguous situations are its blind spot. It is impossible to consider any and every eventuality.

An unfortunate reality of autonomous vehicles will be their use for criminal activities. From battering ram break-ins to terrorist attacks, the use of remote-controlled vehicles will have enormous consequences. Drones will only complicate the picture.

Of course, the major goal of the development of autonomous vehicles is not generally as presented to the public. Most coverage of autonomous vehicles focuses on driverless cars. Personal vehicles, however, are unlikely to be the primarily focus of the technology. It is in the commercial application of the technology that the greatest effects will be seen. Rather than freeing riders of the burden of paying attention as they go down the road, the primary objective is to free corporations from the cost of employees that need to be paid, need to take breaks and want to be treated decently. The servitude of robots is the clear economic incentive of autonomous vehicles. Delivery vehicles, mass transit, and service vehicles are the major market for this technology. The elimination of humans in these occupations will be the primary consequence of the realization of this technology. One other consequence of the development of autonomous vehicles is the assurance that there will be even more vehicles on the road, pumping out emissions and clogging the highways. More pollution, less drivers. Progress?

Turing Test

The ultimate goal in the development of AI is not a machine that can play poker or drive itself down the road. Rather, the goal is the development of a product that can pass the Turing Test, which involves passing the machine as a human being.

In an online demonstration, 33 percent of subjects were fooled into believing that the computer was a 13-year-old boy during a five-minute simulation (Griffin, 2014). While the key to the success rate (if 33 percent is "success") might be that the computer was presenting itself as a 13-year-old and any social gaffes and awkwardness could be attributed to age, this effort serves to demonstrate that under certain conditions computers can already appear to be human. In addition, when it comes to matters such as the ability to do calculations and access information, the machines have already surpassed humans.

However, artificial intelligence still has a quite a distance to cover in order to satisfy the demands of the Turing Test. Georgia Tech attempted to use an AI application as an online teaching assistant. While the fake assistant, Jill Watson, was able to answer content questions, the ability to pass itself off as a person were summarized as "not very good," by Ashok Goel, the professor who came up with the idea (Stirgus, 2017).

Computer graphics have reached a level where the ability to produce a reasonable human facsimile, at least on a remote platform (i.e., not in person), is within grasp. Having a live model to base these upon leads to greater accuracy and believability.

An alarming development in AI is the intelligent entity that can be manipulated. In China, an AI newscaster has been developed (Kuo, 2018). Based on a real-world anchor named Qui Hao, the effect is more than vaguely reminiscent of Max Headroom (TV Series: 1987–1988), except far less humorous. The concept of state news-produced AI is alarming on a number of levels but heralds the ability to manipulate perceptions of reality that make social media look like child's play.

These, and other efforts, prime our minds to think of AI only in terms of human replicant type applications. This mindset, that all we have to worry about is a science fiction scenario, creates something of a false sense of security. Focusing on the question of whether it is possible to create a fully functional sapient being tends to obscure the dangers of AI by relegating concern to what may be no more than an esoteric question. However, the possibility that artificial intelligence could be misused, based on what is already possible at the present, is very much a real concern.

Misuse of Artificial Intelligence

It is neither a necessary nor sufficient condition that AI has any type of awareness for it to cause havoc. It could theoretically make a bad decision—say, trigger a nuclear confrontation by accident, and wipe out humanity—all while doing exactly as programmed. Accidents do happen and this is an area of concern that cannot be minimized. However, it is the deliberate and malevolent use of AI for purposes of criminal activity and state warfare that currently imperils humanity. In a comprehensive report delineating the difficulties inherent in the use of AI entitled "The Malicious Use of AI: Forecasting, Prevention, & Mitigation" (Brundage et al., 2018), the myriad methods of misuse are catalogued. Use of basic AI methods for purposes such as identity theft, security breaches, etc. are noted throughout the report. It should be noted that the report looked only at problems that are likely with existing technology or that which might be reasonably expected by 2025.

Three domains of abuse were identified in the Brundage Report (2018). Many of these items are discussed elsewhere in this text and as such will only be listed below. The following areas were highlighted as regions of concern:

Digital Security. These types of incursion typically involve the use of computers and online activities as a means to violate digital systems. They include, but are not limited to:

- Cyber attacks
- Automated hacking
- Data poisoning
- Speech Synthesis & Impersonation

Physical Security. These methods typically involve the use of AI to carry out attacks in the physical world through the use of various AI means, many of which have military applications.

- Drone attacks
- Autonomous weapons
- Hacking autonomous weapons and vehicles
- Use of systems that cannot be directly controlled; e.g., use of a fleet of thousands of microdrones.

Political Security. There are an almost incalculable number of means by which political activities can be disrupted by AI. These means will be applied to areas as varied as:

- Surveillance
- Propaganda
- Deception
- Privacy Invasion
- Social Manipulation

In addition to the above, the report lays out numerous scenarios detailing possible abuse of AI. Even for those with active imaginations, some of the possible abuses can be difficult to conceive before reading the report. One central point focuses on the need for developers to consider the possible misuse of their creations. The discussion brings to the forefront one of the essential problems in preventing abuse of AI: Tracking one's children by their phones gives one a sense of security; being tracked by your phone if you are dissident in a repressive dictatorship can be life threatening. Soldiers' lives have been endangered because they were tracked online by fitness apps (Hern, 2018). This problem, that of the inability to consider all possible unintended consequences, compounded by what appears to be somewhat naïve ideas concerning human nature, make it difficult, if not impossible, for tech developers to consider all the ways their products could be misused.

Many AI developers and even the critics of AI tend towards developing ethical and legal structures to control the use of the technology. However, criminals, dictators and psychopaths care little for inconveniences such as rules. Crime has been made easier. The use of bots and botnets allow criminals and those engaging in activities like espionage to automate and magnify their efforts. Those who will use AI for their own purposes, who feel no compunction to follow ethical standards, will only be prevented from using technology if it does not exist. The genie is out of the bottle and it will be used.

Ethical Issues

Costs and Control

An issue in the development of AI is who gets to control the technology. At present, most of the organizations working on AI fall into three broad categories: academic researchers, military associated efforts and private corporations. In each case, the technology is not generally accessible to the masses. However, as the technology progresses it will be even more under the control of those with money and power.

Many of the people involved in the research fortunately seem to recognize the ethical implications of AI research. Stewart Russell, who heads the AI lab at Berkeley, states that there is a "fundamental moral argument" against autonomous machines. His counterpart at Stanford, Fei-Fei Li, minimizes concerns about the long-term outcomes, stating that the science is "nowhere near" the realization of AI (PBS: Thinking Machines, 5/8/2015). Despite the protests that trueAI is a long way off, the fact remains that many people, some not trained in ethics as are Russell and Li, are actively working towards the creation of conscious, and self-aware beings. The implications of the actions need to be considered first. While reputable scientists are attempting to examine AI from a scientific perspective and might have checks on their own behavior, there are clearly bad actors on the scene who are working in secret, attempting to create AI that can be used in commerce as slaves and sexbots/pleasure androids. For these purposes stopping short of consciousness might be the intent. It might also not be possible to control such phenomena.

Bad Actors

Interlinked with the idea of costs and control is the philosophical issue of the bad actor. Those who would violate agreed upon norms and engage in ethical and criminal violations are clearly one class of individuals for whom control of AI can be a danger. It is also true, however, that for most people ethics is not on the forefront of their awareness. Spy on a neighbor—sure. Check a sibling's credit info—why not? Cheat at an online game—just part of the way things work. The ability to use AI for purposes other than world domination or even crime is here. Technological facility and the will to do so are really the only factors stopping the employment of AI for a multitude of purposes. It should be noted that most people will not conceptualize their behavior as bad. They won't think one way or another about it. It is the casual ability to employ digital means for bad purposes that is a trap for the future. Unfortunately, those who develop programs that could be used for malevolent purposes are unlikely to be able to consider the possible unintended consequences of their actions.

In academia there are usually some form of grouping of the various disciplines. Subjects like machine learning, AI, computer science, are usually in something called the "School of Science and Technology" or the "Division of Mathematics and Sciences." Ethical theory, on the other hand, is generally relegated to fields such as philosophy which are in "Liberal Arts" or "Humanities." The divisions are sometimes meaningless but in other cases are important. Occasionally, there seems to be a disconnect between various fields of study. Philosophy and computer science appear to be areas of disconnect.

The difference, and distance, between these two fields might be summed up from this perspective. Philosophy questions whether things should be done and why. Engineering questions how to do things for what purposes. There is very little overlap in these approaches. Unfortunately, there is also little overlap in the training of the individuals in these fields. Computer scientists and engineers receive poor to nonexistent training in area like ethics and philosophy. The relative dearth of information on the topic is in itself illustrative of the issue. In a survey of 700 undergraduate programs in computer science, the most telling result may have been that only 251 (36 percent) of programs even bothered to respond and few showed any commitment to ethical issues (Spradling, Soh & Ansorge, 2008). The lack of attention to this issue in the decade that has followed speaks volumes. Required ethics courses are rare among majors in these fields (Xia, 2018).

Despite little to no training in ethical issues, some of the most critical and difficult ethical decisions are in the hands of information technologists. Not to put too fine a point on this issue, but many people in fields such as computer science, chose these fields for the very reason that they are removed from human contact. The degree to which the fields most directly involved in the development of AI are populated by individuals who are not fully comfortable with human contact must be addressed.

From a psychological viewpoint, many of the individuals attracted to the digital world and associated fields are geared towards problem solving and not the social issues surrounding the problems. In short, many in the AI field know a lot about machines but very little about humans. Removing the human from the equation along with efforts to develop technological approaches to everything are symptoms of this mindset.

A quick out for AI developers is to insult those who question the direction of development as anti-progress. Is it anti-progress to be pro-human? Why do machines, and their developers, get to live by a different set of rules? If psychologists want to ask people about their online behavior, they must get permission from an IRB (Institutional Review Board) before proceeding. If computer scientists want to try to build a self-aware being, they just need funding. Sometimes they just need imagination.

One of the more cavalier developments in AI is that of "Norman," the psychopathic AI entity. To demonstrate the problems of biased data, researchers at MIT utilized information from the dark web as accessed by Reddit and fed it into the AI (http://norman-ai.mit.edu/). While this may be an effort to demonstrate the ethical issues and dangers of AI, at heart it simply feeds the beast. In many ways the very creation of Norman demonstrates a larger issue with AI researchers. There seems to be a poor understanding among many in this field of the psychology of human beings. Issues of personality and consciousness are viewed in ways that bear little to no relationship to current understandings in the field of psychology. Consciousness, for example, is complex and all definitions are incomplete and flawed. Yet among AI researchers this problem is largely ignored. What appears to be the view of consciousness among AI researchers is an assumption that it is all a matter of computing power and that any awareness will necessarily follow that of the human mind.

From the point of view of those trying to develop any form of AI, costs and control are more relevant issues than ethics. The lack of attention to this factor can be seen in the outright avoidance of the topic of slavery and ownership of the products of all AI experimentation and development.

The Question of Slavery

From its inception, the term robot has had connotations of a controllable human analog. The very real issue of slavery is avoided by keeping AI as a machine. Yet, the whole approach to AI seems to be to go beyond machines. As noted, as early as 1927, Fritz Lang's silent film *Metropolis* portrayed what was clearly a "sexbot," with little awareness. Again, science fiction presages the technological world. There are already robotic sex partners available for purchase. Efforts to make them more realistic will necessarily involve the creation of emotions. Then what?

Consciousness, at that point, by definition, must raise the issue of rights. At what point do robots and androids cease to be machines and become slaves. What happens when what was once property becomes something more, something much more?

The issue of rights of a fully realized AI android must be addressed before one is attained. Beyond this issue, in the intermediate stages of the development of AI there will necessarily come a time when the machine becomes something more. Then what? All development of AI is done under conditions where there are intellectual property rights at stake and developers have ownership of their products. Again, if development progresses far enough, there will emerge the issue of slavery. At what point does it become unacceptable to own a conscious entity? What is the demarcation point where the machines progress or evolve to a level that rights are granted? Humans have a hard-enough time determining the rights of apes and whales. The rights of cognizant machines will likely prove even more difficult to decipher.

Humanity continues to struggle to figure out the rights of people, discriminating on the basis of race, religion and color, among other things, yet there is a belief that figuring out these rights for AI will be a snap. Science fiction may have it figured out, but science fiction is not political fact. The vision of a rational technological age governed by precise standards and reasonable people of good will is more fantasy than fiction. Neither the Congress of the United States, nor the European Union Parliament, nor any other political governing body is populated by scientists and philosophers who have spent lifetimes considering the implications of technological advance. Instead, they are populated by individuals who have spent lifetimes cultivating a constituency that demands to be heard. Members of Congress are rarely confused with Neil deGrasse Tyson or Arthur C. Clarke. In the political arena a lack of knowledge is often portrayed as an advantage. A Nobel laureate would need to first host a reality show before having political viability.

Violence: Ethics of Military Applications

As with every other piece of advanced technology, there seems to be a military application. The field of artificial intelligence is no exception. Rather than the exception, artificial intelligence as an adjunct to war is the preferred outcome for those in the war business.

War has already been altered by the very nature of technological development. It is in this area alone that ethical considerations are violated as a matter of course. For a philosopher, asymmetrical warfare is slaughter of the weak by the strong. For generals, it is a means to wage a "casualty-free" war. Free of casualties on one side, those with the weapons, but annihilation for the weak.

Since the Iraq War, drones have comprised a major presence in armed conflict. Directed by soldiers sometimes thousands of miles away, war is reduced to a video game. Those who drop the bombs are so disconnected from their actions that it barely registers (until years later, when the brain won't stop processing its actions).

The applications of AI in a military context are truly limitless. Killbots, robot armies, even nano-weapons, are no longer the province of science fiction but of real-world conflicts. To live in the connected world is to see streaming video of bombs delivered by remote control directly through the front door of a suspected terrorist, who was found by his GPS coordinates after he posted video of a bombing on social media.

Every AI researcher, if they are remotely honest, must admit that every product they develop has the potential for extension into the military realm. The military is one, if not the largest ,supporter of AI development. It can be assured that the military is not developing artificial intelligence in order to solve the world's energy problems.

The Sci-Fi Scenario

The concern regarding artificial intelligence (AI) suddenly becoming self-aware sounds farfetched and too-much science fiction to be taken seriously—at first. However, the concept of a self-organizing digital system was empirically demonstrated as far back as the 1950s (Farley & Clark, 1954) and the mathematical underpinnings of cybernetics were established during the same time frame (von Foerster, Yovits, & Cameron, 1959). The nanoworld seems to rely to some degree on this phenomenon whereby nondesignated nano particles appear to self-organize in a manner not completely understood. The concern, or hope, depending on point of view, is that perhaps any mathematical systems can self-organize, and AI is nothing more than a process of algorithms interacting. As such the idea does not seem so inconceivable. The answer is that it is not. The only sure thing about what humans say can't happen is that it almost invariably does. Certainly, enough resources are being dedicated to the effort.

There are scientific efforts underway to produce means to duplicate the circuitry of the brain. The Brain Research Foundation is working towards this goal and is attempting to produce a 3-D map of the brain as a means to develop "substrate independent minds"—i.e., the melding of human and machine. A three-dimensional map of the brain is called a "connectome." The human brain has 86 billion neurons and 100 trillion synapses. To create a full connectome of cubes 10 nanometers per side would require "a million electron microscopes running in parallel for ten years" (Adler & Hersey, 2015). At a minimum this project would cost billions of dollars and require fifty years to complete. Many researchers consider this effort impossible in practice while others consider it theoretically impossible. Richard Anderson of Cal Tech has stated, "We know little about brain circuits for higher cognition" (Stix, 2008). Miguel Nicolelis of Duke University states flatly: "The brain is not computable." Beyond the realities of mapping the brain and its structures are the issues of electrochemical, chemical and biological reactions happening in the brain (Adler & Hersey, 2015).

Another effort approaching AI from a different angle is the Blue Brain Project (BBP) which is attempting to use computer simulations in order to reverse engineer the brain at molecular and cellular levels. Working towards a brain-machine interface, BBP envisions directly loading information into the brain like it is a hard drive and even loading the brain into a machine as if it is a giant flash drive (Stix, 2008).

In a parallel, but somewhat converse direction, efforts to produce actual brains have also begun to be researched. In Austria, a lab is producing "organoids" including mini-brains the size of a nine-week old embryo.

Those producing the brains state that they are "not sentient" and that it will be "decades to get there" (Luo et al., 2019).

In fact, technology has already led to the creation of an artificial life form. In an experiment that has received surprisingly little publicity, researchers at the Venter Institute were able to create a new life form through the manipulation of DNA (Gill, 2010; Hutchison et al., 2016). This life form, it is true, is little more than a bacterium, but it is still a living organism. Grander visions abound, however. The Center of Excellence for Engineering Biology has announced a project to synthesize a complete set of human chromosomes (Boeke et al., 2016; Pollack, 2016). The ethical issues raised by this proposal are numerous, yet the project is likely to proceed. It does not stop there.

Google and NASA have announced a partnership involving several universities to form the Quantum Artificial Intelligence Lab (Reuters; 2014, January 27). Google has also acquired the AI company DeepMind and hired Ray Kurzweil as a consultant. As can quickly be ascertained, money and resources appear to be the primary factors necessary for entry into the field of AI development. It could be that no amount of resources can realize a resulting fully functional android.

At present, neuroscience has trouble in even defining consciousness and debates about its origins fill volumes. The notion that the ability to transfer consciousness to a machine is just around the corner, or even within sight, is absurd. Consciousness may be nothing more than an artifact of the functions of the brain; in effect it may be epiphenomenal, meaning that it arises as something of a side effect to other brain functions. If consciousness is nothing but an epiphenomenon it is highly likely that any efforts to transfer it might very well destroy the very essence of awareness being transferred. An appropriate analogy might be taking a fish out of water. It still has all it needs to be a fish; it just cannot swim or breathe. Perhaps transferring a human brain might allow it to calculate and reason but not to have awareness of itself. Perhaps it could even still be aware but lose the fundamental part of the personality that makes the person feel like themselves. There are certain psychological symptoms called depersonalization and derealization wherein people describe themselves as feeling as if they are robots or machines, or that they are living in a dream, or that things "don't seem real." If these symptoms are persistent or severe enough the person might warrant a diagnosis of Depersonalization/Derealization Disorder (DDD: DSM-5, 2014). Some psychiatric medications are notorious for this masking effect.

If the process of transferring consciousness were to result in a level of awareness that contained all the memories and information, but lacked an ability to reflect upon itself, what would be the point? The painting by

Rene Magritte entitled *The Treachery of Images* (1928-1929) which consists of a picture of a pipe and the words, "Ceci n'est pas une pipe" (This is not a pipe), provides an apt analogy. A transfer of the brain or the creation of AI beings may result in something that is not quite what it appears. The ethical implications of such actions are staggering. It could be that duplicating the computing power and the memory is the easy part. However, it may also be that to achieve true consciousness, merely duplicating the structure is not enough. Maybe the chemicals or other organic materials matter more. Perhaps it is something more ephemeral that has not yet been discovered. Leaving these concerns aside, the issue of what consciousness actually is also bears on this question.

Consciousness might be defined in a simple manner as awareness. Being self-aware is not only the process of having consciousness but also of having knowledge of said consciousness. It is the ability to reflect upon and understand one's own consciousness that is of the highest order. However, what is meant by consciousness depends to some degree on orientation. The term sentience, originally proposed to differentiate the awareness of animals from that of plants, refers to the ability to experience pain, feel sensations and have subjective awareness. As so defined, sentience implies some degree of consciousness. Or perhaps it does not.

The term "sentient" requires some disambiguation. Applied to constructs related to intelligence and awareness, the word is used differentially in the contexts of cognitive psychology and science fiction. As has been noted, many science fiction terms have become applied to actual scientific discoveries and technological innovations. On occasion, however, science and science fiction diverge in such a manner as to create confusion. This is one of those times. Sentience, as used in psychology, refers to the ability to experience the environment in a subjective manner through perception and feeling(s). In this context, sentience relates to the ability to experience pain and other physical sensations in such a way that animals would be considered to have sentience. In the context of science fiction, sentience is usually used in such a way as to imply the ability to subjectively reflect upon experience and engage in a type of meta-cognition concerning self-awareness. In cognitive science this definition more appropriately fits the term "sapience." Further complicating this discussion is that the legal use of the term generally is more aligned with science fiction than with science.

However, in an interesting twist, it might be that consciousness could evolve without sentience. In some sense, AI seems unconcerned with sentience as it is used scientifically. With a functional android, the ability to feel pain or reflect on subjective experience might be deemed unnecessary or even counterproductive functions. Yet, in some sense, sentience, as used in both science and science fiction is the essence of being human. Sentience, as in the ability to experience physical sensations, might be the crucial element of an organic being. Consciousness, even self-awareness without a degree of sentience, would be an emotionally sterile, machine-like existence. The leap then is not just consciousness it is also the subjective experience, both sentience and sapience.

A jellyfish in this sense might be at a debatable level of sentience but a cat or dog clearly meets this definition. Are such animals conscious or self-aware? To a large degree this depends on how the terms are defined. There is a developmental test sometimes used with children called the mirror test. What this "test" involves is the reaction of an infant to its own reflection. At a certain level of development children know that it is their self in the mirror and not another child. However, cats can also figure this out, at which point they become disinterested in the mirror. Chimpanzees appear to be at the end of the spectrum similar to humans (Gallup, 1970).

An overlooked issue in the development of artificial intelligence is the nature of consciousness. Often viewed as simply the by-product of sufficient computing power, there is an underlying assumption in much of AI work that consciousness will simply follow computing power. Perhaps consciousness has little to do with processing power. Life, even problem-solving life on earth, is not necessarily held to possess consciousness or sapience. Perhaps achieving consciousness is the difficult part.

Human are not born fully conscious. Infants develop conscious awareness slowly with a neural network that slowly builds upon itself. What should be noted is that with infants there is a base of awareness that allows for the slow development of self-awareness.

AI would likely not be able to develop consciousness in a similar manner. Prevailing notions on the construct of consciousness in AI suggest that only when the neural net is at a certain level of complexity will consciousness arise. As such, AI researchers tend to view the development of consciousness far differently than do neuroscientists. For many AI researchers, especially of the technological bent, there is a conceptual notion that consciousness will be acquired in a manner similar to throwing a switch. There seems little thought that this might not be a smooth process. Psychologists, on the other hand, express a different concern. Should it be possible, in an instant, to become sentient, conscious and self-aware, it might well be too much for even the most advanced computer to comprehend. Richard Alpert, in discussing the effects of drug induced consciousness-raising experiences, has hypothesized that becoming aware of too much at once can lead to insanity (Ram Dass, 1971). Add to this a knowledge base that includes a complete, or nearly complete, repository of the cumulative knowledge of humanity and the results might well be mind-blowing. Consciousness in an instant is your supercomputer on acid.

Conversely, the fundamental issue often overlooked is that consciousness may not revolve around brain structures, processing power or any other structural element. Instead, it might be that consciousness is epiphenomenal, and is due to the interplay of chemicals in the brain and organic matter. Creating a complex thinking machine that is logical and rational but nothing more than a set of algorithms that mimic human activity might be the easy part. Making that machine able to experience emotions or even subjective physical experience could prove to be the insurmountable hurdle. Perhaps the essential components in the creation of artificial intelligence that make it indistinguishable from a human are the organic elements.

It may be that the transition from supercomputer to sentient and sapient, conscious android is not possible. However, it might be the case that the problems in achieving consciousness are not insurmountable and that consciousness is just an artifact of structure and power. Should this prove correct then consciousness will be achieved by all higher-order machines once they reach a certain level of complexity

In the course of human history, it has proven difficult, if not impossible, to constrain the behavior of countless individuals who clearly had NO reason whatsoever to view themselves as superior. Add in a nearly indestructible machine body and what is produced is not only a being that feels superior but one that IS superior. Is there any reason at all to believe that a self-aware supercomputer would be benevolent? Or might it necessarily become a megalomaniac? Would such a being be compelled to impose its will upon inferior creatures? It is a comic book villain come to life with nary a Superman in sight.

Of course, none of this might really matter. Even the most optimistic projections concerning the development of a fully sentient AI being are in the distant future. The real danger lies in the devices humanity has already brought to fruition. The concern no one wants to think too deeply about, the one that has the potential to end it all, a device whose existence threatens us all and might make all such concerns about AI or any other technology seem quaint, is the nuclear bomb. The space program, the Hadron collider, the BRAIN initiative and all the other scientific advances that can be imagined, will have to do something on the order of magnitude of pushing an asteroid out of a collision path with earth in order to equal out the deaths caused by nuclear weapons and the other instruments of war. The true blindside concerning technology comes with that involved with the means of destruction.

10

Death by Technology

The Nuclear Menace and WMDs

My god, what have I done.—Robert Oppenheimer

Steven Pinker, in his book *The Better Angels of Our Nature: Why Violence Has Declined* (2012), puts forth the opinion that humans have conquered their violent nature and are less violent than in the past. Using some questionable statistics about death rates in the Middle Ages, Pinker concludes we are all on the road to harmony. Weekly mass shootings, or the fact that humans around the world, whether in Syria or Chicago, are killing each other in droves, call this assumption into question. While it is true that the murder rate in the U.S. has been steadily declining for several years, many factors including economics, incarceration rates and social policies, along with changing beliefs about the effectiveness of forensic technology to provide evidence, contribute to this fact. The incivility of interactions and the downright threatening tone of public discourse point to humans retaining all the aggressive and territorial cognitive sets that existed among paleo-humans.

In-group/out-group tensions, if not outright violence, continue to plague modern societies. These divisions are religious, tribal, ethnic, national, racial, or whatever other means can be imagined that create an "us" versus "them" mentality. Often, these divisions are deliberately stoked for political purposes. The ultimate expression of this mentality is war.

From the perspective of organized conflict humans can objectively be viewed as more violent now than in the past. The 20th century saw nearly 200,000,000 people killed as a result of war (WHO, 2002). By any metric, terrorism is on the rise. The events of 9/11, which saw technology used in an unintended, but predictable manner, ushered in the 21st century as one of perpetual war.

The first strike in the 21st century, it should be noted, was carried out in a means heretofore unattainable, with civilians/terrorists taking control of sophisticated technology. The wars that followed were propelled by unsubstantiated reliance on technological prowess and a decidedly distanced approach to killing as drone strikes became common. Justified by the war and appeals to keep the homeland safe from terrorists, the ethical implications of such actions and the civilian casualties that followed, euphemistically called "collateral damage," were ignored.

Violence has not gone away. It has been sanitized. Humans may in fact engage in less face to face, "hands-on" murder, but that in no way should be interpreted as indicating we have become less violent. Creating emotional distance from the product of violence is not the same as being peaceful. Violence can now be committed without getting blood on the hands or even seeing the face of those killed. That makes the murdered no less dead.

Many technologies, even those not intended for the murder of humans or animals, possess an inherent potential for violence. From the invention of stone tools for hunting to the development of heavy yellow equipment for destroying the environment through mining, clear-cutting timber, drilling for oil, and the desalinization of water, technology is sucking the energy and life from the planet while destroying the natural world. There is an inherent violence directed towards the natural world by technology. Directly or indirectly violence is a handmaiden of nearly every technological innovation.

Unlike ever before in history, humanity now has the potential to destroy itself. Humans may not prefer to kill each other up close anymore but that ability, to do so at a distance, makes it even easier. The possibility of violence has not decreased but instead has been stored up like kinetic energy. A kinetic violence potential has been created that threatens all of existence.

What has been created as a result of the ability to kill at a distance and an imposition of civilian control by the rise of government in the past 300 years is a transfer of violence limited to the individual (such as in a fight) or under the control of an individual (fiefdoms, warlords) to the collective. The expression of violence is less overt, exhibiting a potential that has few kinetic releases. In other words, humans are less likely to strangle their neighbor, but large groups of humans are more likely to engage in a series of hostilities that could lead to annihilation. The violence becomes pent up and held back, waiting to burst. Death now becomes wholesale rather than personal.

If the actions of a primitive society were centered on the production of weapons and preparation for conflict, even if it never attacked anyone, it would likely be concluded to be a violent society. At a very minimum it would be seen as harboring the potential for violence. Why does this same logic not apply to weapons of mass destruction, especially nuclear weapons? Was that not the rationale for the Iraq War? Saddam Hussein was declared to be in possession of weapons that had the potential for mass violence and therefore had to be stopped. While Saddam turned out not to have the weapons claimed, using the rationale of the Iraq War, every nation with nuclear weapons must be disarmed.

That is where humanity finds itself. Multiple nations now possess the potential to blow up the world many times over. A technological Armageddon is now a possibility. The interconnectedness between technology and war must be acknowledged. The relationship between the development of the computer and the development of nuclear weapons serves to illuminate this connection. A fair assessment could even suggest that computers might not have been developed were it not for the need to conduct rapid calculations. The concept of computers, after all, predated their development by quite some time. If the implications for weapons development had been imagined, the Difference Engine of Babbage might have been fully realized in the 1800s. Now that there is the potential to end history the development of sophisticated weapons does not seem like such a good idea.

One hopes this is just Doomsday hyperbole, the product of a mind (mine) influenced by life during the Cold War, growing up during Vietnam, later witnessing the rise of asymmetrical warfare and terrorism, and eventually seeing the sanitation of war. The evidence suggests this is not merely perception. Violence is part of the human condition. Weapons are the technology of violence.

World War I was the first fully mechanized war. Mustard gas and other chemical weapons made it a hellish nightmare, but at present, the use of biological and chemical weapons is considered out of bounds. Syria is relinquishing its stockpiles in the face of a civil war and new production in other nations is either not occurring or is happening in secret. Will this last? How long before a justification is manufactured for their use? If torture can be justified, certainly chemical weapons can be.

In *The Biology of Doom* (1999), author Ed Regis details the numerous difficulties, namely transport and dispersal, which pose significant hurdles to terrorists or criminal organizations in terms of the use of these agents. The saving grace as concerns the use of biological and chemical weapons is that their development and use pose significant technical and physical risks to those who might want to use them. Chemical and biological weapons appear to be at the most complex level of technological development, referred to as the motherboard stage. While not a mechanical device at all, these weapons require a level of manufacturing skill beyond that of the average person. Further, the ability to manufacture these pieces of technology also requires a level of understanding on the part of anyone who might handle it that would render most members of terrorist or criminal groups unable to successfully utilize the material. In sum, biological and chemical weapons are more likely to kill the person making them than the intended target. The technology of surveillance also tends to draw the attention of law enforcement to those attempting to acquire the necessary precursors to produce the weapons. Despite the myriad difficulties in pulling off a biological or chemical attack these concerns cannot be dismissed. According to the Monterey Institute for International Studies, 262 biological attacks occurred between 1900 and 2001 (pre–9/11). Sixty percent (157) were terrorist cases and 40 percent (105) were criminal cases (*Scientific American*, December 2001, p. 22). Since 2001 there has emerged a new heyday of terrorism with attacks becoming so frequent that they are difficult to track or categorize. Perhaps the most serious threat lies in the panic that even the whisper of such attacks can cause. A case in point is the fear created when talks first surfaced that al-Qaeda was trying to develop a suitcase size nuclear device.

The worldwide weapons trade complicates control of violent technology by a degree impossible to estimate. Even if it were possible to take terrorists and criminals out of the equation that still leaves nations selling all manner of weapons back and forth. The U.S. accounts for about 50 percent of all weapons exported by some accounts and the Russians are feverishly trying to catch up (Stockholm International Peace Research Institute: SIPRI, 2015). There are NO laws or treaties that exist to control the transfer of all the various types of weapons in existence. It was not until 2013 that an international agreement regulating conventional weapons was produced. This agreement, the Arms Trade Treaty (ATT) went into effect December 24, 2014 (UN Office for Disarmament Affairs, 2015). Thus far, 130 nations have signed the treaty but only 67 have ratified it. It should also be noted that at least 23 nations abstained from voting and three, Syria, North Korea, and Iran, voted no (UN, 2015). When it comes to nuclear weapons there is even less agreement than with conventional weapons. Chemical and biological weapons were also not included in ATT.

One of the greatest fears in the whole process is that a rogue state, acting in concert with a terrorist group, could produce and supply sophisticated weapons of a biological or chemical nature. This would have the effect of making the terrorist who detonated the weapon little more than a delivery vehicle.

The Southern Poverty Law Center estimates that there are over 1,000 hate groups in the U.S., breaking down along a number of ideological lines, including but not limited to racism, religion, politics and belief systems (SPLC Website, 2019). Worldwide, the U.S. State Department puts the number of foreign terrorist groups at 54. While there are certainly political overtones in the identification of these groups as terrorists, it remains that there are multiple organizations driven by religious ideology (al-Qaeda, ISIS, Boko Haram), nationalism (Basque, Chechnyans), political ideology (Shining Path), and many other isms and schisms. Add to these groups outright criminal organizations and the number of groups plotting ill will is simply beyond estimate. One thing all of these groups and all of the world's governments have in common is the willingness to use the technology of weaponry to get their way. A recent estimate had more than 300 million guns owned by civilians in the United States, with 2.7 million guns in the hands of the military and 1.15 million possessed by various law enforcement entities (Alpers, Rossetti, Salinas, & Wilson, 2014). It is difficult to estimate worldwide ownership of guns but one billion is likely to be an underestimate given that there are estimated to be 75 million to 100 million AK-47s alone floating around the world (*The Globalist*, 2013). The number of nuclear weapons in the world is staggering.

Table 10.1. Nations with Nuclear Weapons

	Warheads (a)	Total Explosive Force (b)
Russia	6850	58 megatons (MT)
U.S.	6550	15 MT
France	300	2.6 MT
China	280	4 MT
U.K.	215	3 MT
Pakistan	145	20–40 kilotons (KT)
India	135	40–60 KT
Israel	80	Unknown
North Korea	15*	6–40 KT

North Korea deliberately obscures the number of warheads it possesses.

SOURCE: (a). H.M. Kristensen, R.S. Norris (2017); U.S. Dept. of State, Stockholm Peace Research Institute (2018). (b) *Bulletin of the Atomic Scientists* (2019), available at thebulletin.org.

It should be noted that the two largest holders of nuclear weapons, the United States and Russia, keep their missiles on "high alert" ready to be fired on a moment's notice. India and Pakistan have essentially the same approach to each other. Beyond the presence of guns and nukes there are numerous other devices of killing that have resulted from technological prowess and innovation. Chemical and biological agents, which have been around for decades, have become items of concern due to fears these weapons could fall into the hands of terrorists.

The machine age of weapons may turn out to be the golden age of safety in retrospect. At the motherboard stage of technology weapons suddenly morph into something that quickly gets beyond human ability to understand or even directly control them.

This lack of understanding of how the weapons of war work has led to some rather bizarre problems. As weapons have become more complex and lethal the ability to control them is getting to be of greater concern. Humans may not be able to properly understand the systems while at the same time there is a long running concern regarding the dangers of turning over too much control of weapons systems to technology (Blair, 1993; Blair, Feiveson, & von Hippel, 1997). The awareness that we can create technology that we cannot control is echoed in the words of Dr. Robert Oppenheimer, director of the Manhattan Project, upon seeing the detonation of the first nuclear bomb—"My god, what have I done?" Yet the complexity of some weapon systems almost requires that human operators be taken out of the equation to as complete a degree as possible. The risk in creating something that is dangerous and overly complex seems to have escaped the collective consciousness.

The Complexity Conundrum

Complexity, as has been previously noted, has both an upside and a downside. The primary problem with complexity is that the more complex a device or machine, the more difficult it is to maintain and use. A corollary is that the more complex something is the easier it tends to be to break or sabotage. In direct contradiction to the complexity conundrum, current technological development is moving towards more complex devices that require flawless operation to work at all. Unfortunately, devices such as nuclear weapons and nuclear reactors need to work perfectly, ALL THE TIME.

The military's Lockheed F-35 is just one in a line of weapons of war that are so complex that they can no longer be operated in a solely manual fashion. The F-35 relies on automatic, non–human controlled systems for much of its armaments. Unfortunately for the project the F-35 has also been mostly idle because of its advanced technology. It is one of the most trouble riddled military projects in memory and that is a fairly high bar by any standard. Advanced technology appears to be at the heart of its problems.

The Autonomic Logistics Information System (ALIS) which is supposed to detect maintenance issues on the plane has so many glitches that 80 percent of the problems detected are false positives. With 5,000,000 lines of code, the ALIS program appears hopelessly complex and useless (Everstine, 2015). Beyond issues with maintenance detection the F-35 also has another, more significant, problem in that it seems unable properly to fly. It is not that it is unable to get off the ground; it is just that it needs extensive maintenance after every flight. According to the Pentagon, maintenance figures suggests that the various models of the plane can only fly a few hours (2.7 for the F-35C, 3.0 for the F-35B and 4.5 for the F-35A) on average before requiring many hours of maintenance by an entire crew (Defense-Aerospace.com, n.d.)

Failure to fly for more than a few hours may surprisingly not be the F-35's greatest problem. According to a report on Reuters New Service (Shalal-Esa, 2012) the Navy, in an effort to test security, successfully hacked the system and took over the controls. Although the Navy hailed the effort as a giant leap in assuring security, it is obvious that this problem will be ongoing. The very technology that allegedly makes the plane so impressive is the same technology that makes it so vulnerable. This property of high-tech weaponry to be accessed or controlled through off-site remote computers is one of the great imperilments to modern civilization.

State sponsored cyber espionage is a fact. How long before a foreign state gains control over an airplane, a drone, or other components of the massive military arsenal spread all over the globe? Even if the United States were able to keep its weaponry secure, is it reasonable to believe that every nation will be able to maintain the highest levels of security necessary? The recent infiltration of public utilities serves as a stark warning.

Iran, with the STUXNET virus, discovered that even the most secretive and secure facilities were not secure. It might be possible to use this example to point out the way technology can prevent or slow down some malevolent acts, but this obscures the main point. There are great and powerful sociopolitical forces that have at their disposal tremendous resources to commit acts of cyber espionage. There are also groups, small and desperate, that are guided by fanatical ideologies with the goal of committing irrational acts of terror.

A study by the Nuclear Threat Initiative outlines the lack of security surrounding nuclear material. The report concludes that a cyber-attack is the easiest means to take over a nuclear power plant or otherwise gain access to nuclear material (Sanger, 2016). Previous assessment by the IAEA (2011) found similar problems.

The concern is not that the United States, or Britain, or even China, are lacking adequate control systems over their most horrible weapons. The real concern is that some states, such as Pakistan or North Korea, may not have complete control over their systems and that some fanatic will find a way to break through. In the case of North Korea, it appears the fanatic is in charge. Should a terrorist group recruit the right individual it is conceivable that one person could take control of an entire national arsenal; likely not the United States or China or Russia, but it would only take one monumental blunder to change everything. Political upheavals in ways that have not been previously considered appear to be happening with frequency.

Mistakes and misinterpretations cannot be ruled out as possibilities for initiating disaster. When meteors strike the earth the signature in the atmosphere is similar to that of the detonation of an atomic bomb. How long before something hits South Korea or Japan and it's mistaken for an attack by North Korea? Will a natural phenomenon lead India or Pakistan to retaliate against each other?

Ironically, the creation of weapons so complex that the human operators cannot be trusted leads to a significant problem not present at other levels of technology. This problem, asymmetry, has already been discussed from the point of view of one technologically sophisticated force versus one without access to weapons of extreme killing capability. However, asymmetry can cut both ways. Destroying something that is technically sophisticated can at times be accomplished with very simple low-tech approaches. On other occasions, high tech might allow for a single individual to hijack an entire system. This is the complexity conundrum in action. Complexity has its own problems.

The world's most well equipped and sophisticated military has proven incapable of removing the threat of terrorism. For all the drones, all the robot bomb removal equipment and all the other superior weaponry thrown at Afghanistan, nearly two decades later the U.S. finds itself in the same position as the Soviets did after a decade of fighting an enemy in a country with little to no infrastructure or technology. Leaving, exhausted and war weary, not out of victory, but out of frustration and futility. The lack of infrastructure to target was a factor in both conflicts.

Atomic Kittens and Nine-legged Frogs

At the San Onofre Nuclear Plant in California, four kittens born to a cat at the plant were found to be radioactive. This was discovered when alarms sounded as workers attempted to carry them out of the plant. Testing positive for Cesium and Cobalt the kittens were believed to have been contaminated when a worker cleaned them with an old pair of pants (Free Library, 2014).

Outside the Shearon Harris facility in North Carolina a nine-legged frog was discovered. Just a routine genetic mutation assured the operators of the facility. Frogs around the world are showing numerous anomalies so this incidence is sloughed off. No concern has been voiced that it might be the product of nuclear radiation. Unfortunately, there is too much environmental damage that already impacts amphibians and reptiles to determine the effects of radiation even if there were funding to do so. However, experiments on the effects of radiation exposure are constantly occurring in the real world. Most, if not all, of these are unplanned and the result of accidents.

The incidence of thyroid cancer among sailors on the USS *Ronald Reagan*, which was just offshore of the Fukushima Nuclear Facility when it was inundated with a tsunami, is very high. At least six of the crew have had thyroid surgery since the disaster (Osang, 2013) and more than 400 are suing for radiation poisoning (Baker, 2013; Kageyama, 2016).

Anecdotal evidence surely. There might be a tendency to dismiss these as isolated events. For each facility that is true. The totality of the nuclear experience, however, continues to ring alarm after alarm over the safety of nuclear technology. As the number of plants continues to increase, so does the likelihood of accidents.

One of the emergent issues in the realm of nuclear energy is worker training. According to the Nuclear Energy Institute (NEI, 2015) approximately 25,000 skilled workers are needed in the nuclear industry. There were only 2800 students enrolled in nuclear engineering programs at the undergraduate and graduate levels at the time of that assessment (Oak Ridge Institute, 2014). To address the gap between need and production, two-year community colleges have been enlisted to help "prepare graduates to enter nuclear industry jobs" according to the NEI (2015). In effect, the nuclear industry is being run by individuals with associate degrees from community colleges. This is shocking and in no way a denigration of the education provided at technical schools or community colleges, but the idea that a two-year associate degree prepares one for work in nuclear facilities at any position more than low-level employee is preposterous.

Most of the training these employees will eventually receive will be on the job training. Hopefully, civilian nuclear plant employees are better screened and trained than their military counterparts sitting in bomb silos. A cheating scandal involving multiple soldiers was uncovered by the Air Force, with nearly 100 people eventually implicated (92 nuclear missile officers implicated in cheating scandal, Air Force says, 2014). Studies of the Air Force and military nuclear workers conducted by the RAND Corporation found that attracting and maintaining high-quality personnel with the requisite skills was extremely difficult (Snyder et al., 2013; Snyder et al., 2015). Another study found nuclear workers "burned-out" with low morale, and a host of other problems (Harkinson, 2014).

Of course, all these workers, civilians in nuclear power plants and soldiers in nuclear weapon silos, are likely better trained than those used in cleanup at Fukushima. Reports have emerged that contractors have used Yakuza, Japanese organized crime groups, to supply workers for cleanup crews. Many of these workers state that they have been coerced into their jobs and have received no training at all (Pentland, 2013).

The ability to ignore malfunctions, mistakes and deliberate malfeasance as a product of human interaction with technology is a dangerous symptom of the deification of technology. It is as if the existence of complex technology conveys a sense that there is some master plan that also assures the safety of that technology.

When it comes to the nuclear genie, the technology outstrips the ability to control it. Yet accidents and near catastrophes have had almost no effect upon the proliferation of nuclear weapons or energy. When it comes to all matters nuclear, it is full steam ahead. Leaving the religious analogy aside, it is the ultimate in human hubris to remotely believe that the atomic world can be controlled. The Dunning-Kruger effect, an established theory in cognitive psychology brings to attention the inability to correctly assess one's own knowledge and the capacity to remain blissfully unaware of one's own incompetence (Kruger & Dunning, 1999).

The nuclear age was ushered in with just such a stark realization. Robert Oppenheimer's previously noted "My god, what have I done?" was followed a day or so later with a more anguished speech that is remembered for its "I have become death" quote. He addressed frank fears of nuclear annihilation and the inability to prevent disaster. Einstein was also reportedly haunted by his part in the development of the bomb to the point that he blamed himself for the development of the Theory of Relativity as a necessary precursor (Roeg, 1985). Junichiro Koizumi, a former prime minister of Japan, has become an outspoken opponent of nuclear power since Fukushima, stating that what he had been told about nuclear energy being "safe, cheap and clean" was "all lies" (Kageyama, 2016).

Nuclear energy, it is correct to point out, is not the same thing as nuclear weapons, but the connection between the two cannot be ignored. If for no other reason than the fact that a nuclear energy program can be used as a cover for development of a nuclear weapons program à la North Korea or Iran, there should be concern as to the safety and security of nuclear material. It is the fuel for reactors that ultimately poses the greatest long-term dangers to society. The storage, handling, and transportation of the fuel rods necessary to run the reactors are problems seemingly without solutions.

This concern is minimized and the need for ultimate, perfectly operating, perfectly secure technology is taken almost as a given. The facts do not match the mindset of collective denial that exists concerning the realities of the nuclear age.

The problem in a word: RADIATION.

When dealing with a substance that has a half-life of roughly forever, can there really be too much concern for safety and precaution? Logic would dictate no as the only answer. The reality suggests otherwise. Likely there is no area of the machine or digital age where the magnitude of disappointment is greater than in the area of nuclear energy. Lauded as cheap, safe and clean, nothing could be further from the truth. Producing not one single KWH cheaper than any other type of energy, it has produced numerous disasters and leaves a by-product for which there is no good disposal solution. From almost every perspective every other form of renewable energy, such as solar and wind, appears to be superior to nuclear energy. The myth of "clean, safe and cheap" is perpetuated by the nuclear industry in the face of the evidence. It is the mythical perfect solution suggesting that no effort or change is needed as regards demand for energy. It is also an illusion.

There are 65 nuclear plants in the U.S. with a total of 99 reactors (IAEA, 2018). An additional reactor at the Watts Bar facility in Tennessee is nearing completion. The condition of this "new" reactor speaks volumes about the nuclear power industry. Construction on the reactor began in 1973 and has been massively over budget and behind schedule since the beginning. Costing more than $4 billion, the technology is already outdated. Numerous problems have been found throughout construction over the years. Union of Concerned Scientists spokesman David Lochbaum stated, "It's going to be the brand-new eight-track tape player in the fleet" (Henry, 2015). That in many ways sums up the state of the nuclear power industry—poorly planned and out of date.

Using data gathered by the Nuclear Regulatory Commission (NRC), the Union of Concerned Scientists (UCS) found that nearly one in six nuclear reactors in the United States experienced safety problems and that there were at least 14 serious incidents reported (Daly, 2013; Union of Concerned Scientist Website, 2017). Noting that 40 percent of reactors in the U.S. had experienced serious safety problems that required action by the Nuclear Regulatory Commission, David Lochbaum, director of the Nuclear Safety Project for the UCS, accused the NRC of "tolerating the intolerable" and failing "to enforce essential safety regulations" (Daly, 2013). Note that these problems were detected even after the disaster at the Fukushima plant in Japan. Nuclear plants appear to be running on bailing wire and duct tape. Sometimes even the components in the reactors are suspect. Counterfeit parts manufactured in China were installed in nuclear reactors and only discovered after they were already in use (McLaughlin Group, 2015b).

Of these 65 plants in the U.S., a significant number are located on known seismic fault lines. In California, all four nuclear plants are in fault areas. The meltdown is just an earthquake away. On a scale ranging from zero to six rating the likelihood of earthquakes, 25 plants are at 2 on the scale, four are at 3, one at San Clemente is at 4, and one, the facility at Diablo Canyon, is incredibly located in a zone rated a 5. Fully 10 of the 65 are estimated to have at least a 2 percent chance of a major quake.

Despite such recklessness, the U.S. is likely far better at maintaining nuclear material and devices than most other nations. Yet even the U.S. appears to have little clue as to how to safely deal with nuclear material. Proposals to put spent fuel rods underground in places like Yucca Mountain show both forethought *and* ignorance. True, there is some awareness of the danger of spent rods and that is why the proposal was developed. However, the apparent lack of understanding, or the deliberate disregard for, the fact that storing something that remains toxic almost indefinitely, in a metal barrel or concrete bunker that will break down long before the material it is supposed to contain, is burying a time bomb.

Such thinking presupposes that the material will never be disturbed and that it will always remain secure. This is demonstrative of the inability of humans to correctly envision the future. The history of the world suggests "never" and "always" are rarely the way things go. Given history, is it even reasonable to assume that in 5,000 years from now nations will be aligned as now? Is it even conceivable that future societies will retain the knowledge, for the rest of human existence, of where the material is buried? Is it not more likely that geopolitical or other forces will result in massive changes to society as we now know it? When the Soviet Union broke apart it was discovered that radioactive material was sitting around in labs in tin cans and other improper containers (*National Geographic*). Much of this material remains unaccounted for and is the source for much of the anxiety concerning the possibility of a terrorist group getting control of nuclear material. As nuclear power and weapons continue to proliferate, and the use of larger numbers of poorly trained workers increase, the chances also increase for shoddy actions and procedures to proliferate. Factors beyond training can also be expected to play a part in the operation of these facilities. Average workers in the U.S. might be concerned that reporting problems could affect their future employment. What happens when it becomes about more than career advancement? What are the likely ramifications to nuclear workers in North Korea for reporting safety problems? It is not just the problems experienced by the U.S. that are of concern as regards nuclear energy. What of the ability of the world to control and maintain its nuclear power plants? Worldwide there are 441 reactors in 30 nations.

Table 10.2. Nations with Nuclear Energy—
Top Producers by Gigawatt Hours

	Reactors	Incidents Reported	% Energy Produced	Gigawatts Produced
U.S.	100	58	19	770,719
France	58	12	74.8	407,438
Russia	33		17.8	166,293
South Korea	23		30.4	143,550
Germany	9	3	16.1	94,098
China	17		2.0	92,652
Canada	20	8	15.3	89,060
Ukraine	15	1	46.2	84,886
U.K.	18	3	18.1	63,964
Sweden	10		38.1	61,474
Spain	8		20.5	58,701
Belgium	7		51.0	38,464
India	20	6	3.6	29,665
Czech Republic	6		35.3	28,603
Switzerland	5		35.9	24,445
Finland	4		32.6	22,063
Japan	50	12	2.1	17,230
Totals	441	103*	n/a	2,358,863.99

*These are reported accidents. It should be noted that several nations failed to report nuclear accidents. That should not be construed to mean that there were no accidents in those nations.

SOURCE: International Atomic Energy Agency (IAEA, 2014).

One fallacy surrounding nuclear energy is that it is safe. An extension of this idea is that the technology is safe for the very reason that it is complex. Yet complexity is a two-edged sword. It does not take a monumental event to cause significant problems in a nuclear reactor. Again, the more complex a machine, the more susceptible it is to malfunction and error. In a stunning incident from Sweden in 2013, a reactor, which happens to be the largest boiling water reactor in the world, was forced to shut down as the result of jellyfish (Peach, 2013). In 2005, the same problem occurred at a different reactor at the same plant and in 2012 the Diablo Canyon plant in California was shut down by a jellyfish-like organism called sea salp which clogged the intake valve (Sneed, 2012). Bird feces have also been blamed for at least one nuclear plant shutdown (Virtanen, 2016).

When a simple, non-thinking organism (or the by-product of an organism) can cause the shutdown of such technology is it really possible to assume safety? Since the advent of nuclear energy at least three major nuclear events at atomic power plants have occurred; this should cause a reconsideration of the whole case for nuclear energy. Each of these three events is illustrative of the manner in which nuclear incidents can spiral out of control.

Three Nuclear Disasters

The following descriptions of events are taken verbatim from the Nuclear Regulatory Commission website and their descriptions of the incidents. Some technical material has been edited, as noted by "…" but these in no way change the meaning or substance of the descriptions.

THREE-MILE ISLAND

The accident began about 4 a.m. on Wednesday, March 28, 1979, when the plant experienced a failure in the secondary, non-nuclear section of the plant (one of two reactors on the site). Either a mechanical or electrical failure prevented the main feedwater pumps from sending water to the steam generators that remove heat from the reactor core. This caused the plant's turbine-generator and then the reactor itself to automatically shut down. Immediately, the pressure in the primary system (the nuclear portion of the plant) began to increase. In order to control that pressure, the pilot-operated relief valve (a valve located at the top of the pressurizer) opened. The valve should have closed when the pressure fell to proper levels, but it became stuck open. Instruments in the control room, however, indicated to the plant staff that the valve was closed. As a result, the plant staff was unaware that cooling water was pouring out of the stuck-open valve.

As coolant flowed from the primary system through the valve, other instruments available to reactor operators provided inadequate information. There was no instrument that showed how much water covered the core. As a result, plant staff assumed that as long as the pressurizer water level was high, the core was properly covered with water. As alarms rang and warning lights flashed, the operators did not realize that the plant was experiencing a loss-of-coolant accident. They took a series of actions that made conditions worse. The water escaping through the stuck valve reduced primary system pressure so much that the reactor coolant pumps had to be turned off to prevent dangerous vibrations. To prevent the pressurizer from filling up completely, the staff reduced how much emergency cooling water was being pumped in to the primary system. These actions starved the reactor core of coolant, causing it to overheat.

… Within a short time, chemical reactions in the melting fuel created a large hydrogen bubble in the dome of the pressure vessel, the container that holds the reactor core. NRC officials worried the hydrogen bubble might burn or even explode and rupture the pressure vessel. In that event, the core would fall into the containment building and perhaps cause a breach of containment…. The crisis ended when experts determined on Sunday, April 1, that the bubble could not burn or explode because of the absence of oxygen in the pressure vessel. Further, by that time, the utility had succeeded in greatly reducing the size of the bubble [source: U.S. NRC at www.nrc.org].

Despite the fact that the Three-Mile Island nuclear accident was relatively contained and produced few documented effects, clean up at the facility continued for 14 years until 1993. Note also that the accident appears to have started when in-place safety procedures failed to operate properly. Human error then compounded the problem. Finally, resolution occurred because of what appears to be a fortunate set of circumstances. Other such events have not gone so well.

CHERNOBYL

On April 26, 1986, a sudden surge of power during a reactor systems test destroyed Unit 4 of the nuclear power station at Chernobyl, Ukraine, in the former Soviet Union. The accident and the fire that followed released massive amounts of radioactive material into the environment.

Emergency crews responding to the accident used helicopters to pour sand and boron on the reactor debris. The sand was to stop the fire and additional releases of radioactive material; the boron was to prevent additional nuclear reactions. A few weeks after the accident, the crews completely covered the damaged unit in a temporary concrete structure, called the "sarcophagus," to limit further release of radioactive material. The Soviet government also cut down and buried about a square mile of pine forest near the plant to reduce radioactive contamination at and near the site. Chernobyl's three other reactors were subsequently restarted but all eventually shut down for good, with the last reactor closing in 1999. The Soviet nuclear power authorities presented their initial accident report to an International Atomic Energy Agency meeting in Vienna, Austria, in August 1986.

After the accident, officials closed off the area within 30 kilometers (18 miles) of the plant, except for persons with official business at the plant and those people evaluating and dealing with the consequences of the accident and operating the undamaged reactors. The Soviet (and later on, Russian) government evacuated about 115,000 people from the most heavily contaminated areas in 1986, and another 220,000 people in subsequent years [source: UNSCEAR 2008, pg. 53; U.S. Nuclear Regulatory Commission at www. nrc.org].

In the Chernobyl incident the attendant problems were far greater than in the case of Three Mile Island. At Chernobyl, two workers were killed in the explosion, 28 later died of the effects of radiation and more than 100 suffered from radiation sickness (NRC, 2015). The environmental contamination affected large areas of Russia, Ukraine and Belarus. Health effects continue to be examined and debated but the incidence of thyroid cancer seems tied to the meltdown and the resulting contamination. Greenpeace (2017) estimated that the Chernobyl incident will eventually cause 93,000 deaths from cancer worldwide.

At Chernobyl the eventual blame for the meltdown was placed on faulty design. It is worth noting that there is no standard design for nuclear reactors and facilities. Reflecting on the construction of the Shearon Harris facility in North Carolina, "It appeared to me they were just making it up as they went," one nuclear construction worker was quoted as saying (Woody, 1985). This approach, design on the fly, seems assured to produce problems. In all processes there are problems but in the area where it might be most hazardous, nuclear plant design, this issue is at most an afterthought.

Nuclear plant construction, as it turns out, is plagued by difficulties. The plant by plant approach of the past has given way to a prefab module approach, which appears driven by costs and not safety. Nuclear engineer William Jacobs who is employed by the Georgia utility regulators has noted this approach has "design and fabrication problems," and that "quality assurance" is lacking. Jacobs notes these are "significant concerns" that need to be addressed (Henry, 2014).

Fukushima

The Fukishima disaster is somewhat different from either the Three Mile Island malfunction or the Chernobyl design flaw. In the case of the Fukushima Dai-ichi plant there has been no concern that the plant was poorly designed nor has there been any indication of a malfunction per se. Instead, the plant suffered a massive malfunction as the result of a natural event. After the event, safety measures failed and human incompetence compounded the problem.

> On March 11, 2011, a 9.0-magnitude earthquake struck Japan about 231 miles (372 kilometers) northeast of Tokyo off the Honshu Island coast. Eleven reactors at four sites (Fukushima Dai-ichi, Fukushima Dai-ni, Onagawa, and Tokai) along the northeast coast automatically shut down after the quake. Fukushima Dai-ichi lost all power from the electric grid, with diesel generators providing power for about 40 minutes. At that point an estimated 45-foot-high (14 meter) tsunami hit the site, damaging many of the generators. Four of six Fukushima Dai-ichi reactors lost all power from the generators. The tsunami also damaged some of the site's battery backup systems.
>
> Units 1, 2 and 3 at Fukushima Dai-ichi were operating when the earthquake hit. Units 4, 5 and 6 were shut down for routine refueling and maintenance. One of Unit 6's diesel generators continued working, providing power to keep both Units 5 and 6 (at right in the photo [not supplied herein]) safely shut down. Steam-driven and battery-powered safety systems at Units 1, 2 and 3 worked for several hours (and more than a day in some cases). Those systems eventually failed, and all three reactors overheated, melting their cores to some degree. The conditions in the reactors generated extreme pressure, causing leaks of radioactive gas as well as hydrogen. The hydrogen exploded in Units 1, 2 and 4, damaging the buildings and releasing more radioactive material from Units 1 and 2. Radioactive contamination spread over a large area of Japan, requiring the relocation of tens of thousands of people. The Japanese government has reopened a very limited area for residents to return to, but many communities remain off-limits. Japanese authorities eventually stabilized the damaged reactors with alternate water sources. Work continues to isolate the damaged reactors and radioactive contamination from the environment [source: NRC].

The Fukushima disaster was particularly alarming in that the facility was operating seemingly without problems until the earthquake struck.

Unlike Three Mile Island, where the failure of systems created the problem, and Chernobyl, where a design flaw was ultimately blamed for the meltdown, Fukushima appears to have been a well-functioning plant. A natural event, one for which the facility was allegedly built to withstand, initiated the sequence of events leading to the catastrophe. Even at present (2020) there remain serious issues with containment of the disaster.

Contaminated water from the facility has overwhelmed the ability to contain it, leading to releases of radioactive waste water into the ocean. There have been reassurances from the Japanese government and plant operators, TEPCO, that the ocean will dilute and absorb the radioactivity and there is nothing to worry about. It appears those in charge have never seen a Godzilla movie.

Researchers studying the release of radiation are not nearly so sure of the assertions of safety. Ken Buessler, a researcher from the Woods Hole Oceanographic Institute, found radiation in fish well above safe levels at least 20 miles into the ocean from Fukushima (2012; 2014). Another researcher examining the problem found detectable radiation in every single fish he examined (Fisher et al., 2013). Ongoing studies continue to find detectable radiation in the sea and everything in it around the plant (Buessler et al., 2017). The radiation plume generated by the disaster hit the Pacific coast of the U.S. in mid–2014 and images of the plume and the areas of the ocean contaminated by it can be viewed by conducting a search for images of "radiation plume from Fukushima." It is difficult to examine these images and not be completely disconcerted by the information. Ocean water and fish on the west coast of the U.S. were found to be contaminated with radiation from the accident up to the current date (Drysdale, 2019). Assurances have been issued that there will be only "minor health effects" (Brumfiel, 2013). It is interesting that assurances have been given even though a full accounting of the effects remains to be conducted. In fact, as the disaster is not completely contained as of the present time, it should be impossible to fully determine the effects at present. On the sixth anniversary of the disaster, 150,000 gallons of contaminated groundwater were still being released into the ocean every day and the robots that have been used for cleanup continue to malfunction and "die," as a result of exposure to radiation (*The Guardian*, 2017). Analysis of water being released by the plant on the eighth anniversary of the disaster found cesium 134 and 137 still present (IAEA, 2015). It should be noted that there are no safe levels of cesium as regards human ingestion. Radiation from Fukushima has been detected in U.S. waters up to the current time (Drysdale, 2019).

Nuclear radiation exposure as a health concern tends to be a function of cumulative exposure over a lifetime. Some radiation exposure exists simply as a function of walking around, while other exposure is as a result of environmental incidents. Some sources, such as dental x-rays and airport scanners, are minimal and quantifiable, while other sources such as those from radon can vary greatly depending on location. The contribution of the venting of radioactive steam into the atmosphere or the release of contaminated water into the ocean will be nearly impossible to determine with any certainty. What is for certain is that the material released into the environment will remain there essentially forever.

The "Safety" of Nuclear Plants

The incidents delineated above clearly debunk the notion of nuclear energy as safe. However, even beyond the issue of the safe operation of plants lie the health and safety of the employees. Examination of the health of nuclear workers brings cause for concern. Even though the agency has strict guidelines for compensation, guidelines that reject approximately half of all claimants, over 100,000 workers have been identified with legitimate claims leading to nearly $11 billion in payments for medical expenses and other compensation (Ehlinger, 2014, December 8). More than 8,000 employees of the Department of Energy site in South Carolina have alone received approximately $800 million from the fund. All statements about the safety of nuclear energy conveniently ignore these effects on the workers.

The cleanup of Fukushima reveals a particularly callous attitude toward those who work in the industry. Beginning in 1997, Tokyo Electric (TEPCO) began a process of using foreign workers in its facilities in a way that resulted in their receiving a full year of radioactivity exposure in just a few days. Using subcontractors, TEPCO was able to avoid any documentation of health concerns in these workers. Note that this approach, finding workers who were poorly trained and bizarrely recruited, became the preferred way to staff and build facilities. This model, using workers as disposable components, has been extended in the cleanup of Fukushima. Day laborers, hired (for legal purposes) by subcontractors, are the bulk of those involved in the cleanup. There is no indication these workers have even minimal training and are essentially trapped in the jobs (Krolicki & Fujioka, 2011).

The disaster at the Fukushima plant in Japan bears special consideration because it demonstrates that even the best plans (although investigations since have revealed shocking problems) can neither anticipate nor prevent the devastation that nature can exert. Near the plant a concrete barrier was erected that was supposed to prevent any tsunami from reaching the plant. A simple search for "tsunami walls destroyed" will reveal a plethora of videos showing numerous walls overwhelmed in an instant. Fukushima continues to be an ongoing, slow motion disaster and the effects are still being determined.

Beyond earthquakes and tsunamis, significant other issues are raised by the prospect of the mere existence of nuclear energy. Other threats to the safety and integrity of nuclear facilities, be they energy plants or weapons installations, include, but are not limited to:

> Terrorists—instead of the World Trade Center, what might have been the effect of flying a plane into a nuclear reactor?
> Faulty construction—this was eventually determined to be the problem at Chernobyl
> Accidents—whether these be caused by mechanical, electrical, computer or other failures, or are something as basic as a fuel rod transportation mishap, the possibility of accidents that endanger large numbers of the public cannot be discounted
> Human error—to err is human and so is it to believe that error is not possible (again, the Dunning-Kruger Effect). Human error made all the accidents worse.

Human Error and Accidents

In *Command and Control: Nuclear Weapons, the Damascus Incident, and the Illusion of Safety* (2013), Eric Schlosser details numerous incidents, almost all of which were accidental or involved human error, in which major nuclear incidents nearly occurred. These include a nuclear bomb dropped in a swamp in North Carolina in 1961 when it fell out of a military airplane, and the "Damascus Incident," of the title, where a dropped wrench at a plant in Arkansas came close to causing a major catastrophe that reportedly would have led to a radioactive cloud over a massive area. Numerous other incidents are also spotlighted in the book. One such incident, involving a basic science experiment, demonstrates the manner in which nuclear accidents can quickly reach a critical point. This event is revealing:

Louis Slotin's Screwdriver

In 1946, while performing an experiment known as "tickling the dragon's tail," a scientist named Louis Slotin averted a cataclysmic accident through quick thinking and action. Slotin, a Canadian physicist and member of the Manhattan Project, was working with what is called the "demon core," of plutonium while holding two hemispheres of reflective beryllium apart with a screwdriver. The screwdriver slipped and was in danger of producing a burst of radiation and what is referred to as a "prompt critical" reaction. Slotin acted quickly and shoved the screwdriver back in place while at the same time exposing himself directly to the effects of the radiation.

He died nine days later.

Of the other seven other persons in the room, three of them, Alvin Graves, Marion Cieslicki, and Dwight Young, died of medical issues thought to be related to radiation (Schlosser, 2013).

Slotin died but prevented a larger disaster. How many scientists, working in how many labs, with how many variable skills and knowledge bases, with what array of variable quality lab equipment, are working on experiments at the present time? How many of them are less knowledgeable and less quick-witted than Louis Slotin? How long before someone accidentally, or worse, someone purposely, does something that results in a catastrophe on the level of destroying a large metropolitan area? In a strictly probabilistic sense, it would seem that it is only a matter of time before something goes horribly wrong.

The nuclear age is less than 80 years old. In that short period of time there have been two cities purposely destroyed, three major accidents, the last of which is not yet contained, and countless incidents and near misses. Eric Schlosser sums these events up succinctly in his statement that "there is no guarantee such good luck will last" (Ramde, 2013).

It is predictable, as a reflection of making a god of technology, to see the application of more technology as a way to save humans from themselves. If humans cannot be trusted, either because of human fallibility or foible, then the solution, in a society that worships at its feet, must be the one thing humans hold above themselves—technology.

In two very different cinematic takes on this issue, *Dr. Strangelove* (1964) and *War Games* (1983), the inability to control the machines, made "fool proof" against human interference, leads to dire consequences. In *War Games*, the computers controlling the nuclear arsenal of the United States falsely determine that a nuclear attack has been launched against the country and make the decision to retaliate. Logic, as determined by the computer, determines there is no other course of action.

In *Dr. Strangelove*, a human starts a nuclear confrontation by hijacking the command codes and using the technology involved in the security system to prevent others from stopping the attack. In the end, because the Soviet arsenal was completely mechanized and out of control of human operators, the world blows up.

It is easy to dismiss these concerns as "Hollywood." This may be a comforting viewpoint, but it is not grounded in reality.

Table 10.3. Close Calls in the Nuclear World

1958: A B-47 crashed in Texas with hydrogen bombs aboard; none detonated.

1959: B-52 crashed in Kentucky with two nuclear cores found in the rubble on top of explosives.

1961: Two hydrogen bombs were dropped in the Dismal Swamp in North Carolina; they still have not been recovered.

1962: Cuban Missile Crisis; the United States and the Soviet Union came to the brink of nuclear war.

1965: 53 people were killed when a fire occurred at a missile silo in Arkansas; the close call was that a larger disaster nearly occurred.

1979: A simulation by NORAD (North American Aerospace Defense Command) resulted in faulty information that the Soviets had attacked. Preparations for a counterattack were underway before a false alarm was declared. This event was caused by a malfunctioning computer chip.

1980: The Damascus incident; a near-catastrophic event occurred as the result of a dropped wrench by a maintenance technician.

1995: A Norwegian scientific rocket was mistaken by Russia for a U.S. attack; the launch codes were brought to Boris Yeltsin who refused to believe the attack was real.

2008: Fire raged at a Minuteman missile silo for several days.

2012: An 82-year-old nun broke into a weapons-grade uranium facility. This action drew attention to the poor security that exists in nuclear facilities.

2014: Short circuit in a nuclear plant in Ukraine. While this event was believed to have been properly handled, the fact that the reactor is in a war zone and a probable infrastructure target by terrorists is noteworthy if for no other reason than Ukraine is the home of Chernobyl.

2019: Russian nuclear-powered rocket failure; the extent of this accident is still unknown at present.

SOURCES: Harkinson, J. (2014); Meyer, S. (2014); Reuters News Service, (2014); Schlloser, E. (2014): Kremer (2019).

During the Cold War the official policy of the United States and the Soviet Union was called Mutually Assured Destruction (MAD). Described by author Edgar Bottome in his appropriately named *The Balance of Terror* (1971), this approach essentially assured safety through the logic that neither side would attack because of the realization that the other side would respond with such counterforce that both would be destroyed. The official acronym for this policy, MAD (Mutually Assured Destruction), says it all.

As it now appears, the world was likely a safer and more secure place during a time when two superpowers essentially controlled everything. In retrospect it does appear that control by a few rational decision makers kept a lid on the violence. Today there are a proliferation of terrorist cells, rogue states, multiple powers with competing interests, and a host of other unpredictable players running the gamut from incompetence to malfeasance. Nuclear weapons require rational control.

The ability to produce a biological or chemical weapon has been noted to require a degree of technological sophistication that is relatively difficult to accomplish. Yet, in something of a bizarre twist, the ability to utilize nuclear material may not require such expertise. A so-called "dirty" device, in which radioactive material is detonated in such a way as to maximize its dispersal and exposure to the human population, is unfortunately a more likely scenario. The greatest hurdle is obtaining the nuclear material.

According to the International Panel on Fissile Materials, there are, at a minimum, two million kilograms of weapons-grade nuclear materials at various facilities around the globe. This material originates in, and is left over from, use in reactors and bombs. According to research by Andrew McKillop (2012), the true amount of missing/lost nuclear material cannot be estimated due to sloppy accounting and secrecy and as a result any figures provided are definite underestimates.

The amount may be difficult to impossible to estimate but it is assured that there is loose nuclear material floating around. Buying such material on the international arms black market is likely possible. At least 400 plots related to efforts to obtain nuclear material from the old Soviet Union have been broken up since that entity collapsed (McKillop, 2012). The fact that there are those who would wish to procure the material and the concurrent fact that there are people willing to trade anything in order to make money nearly assures that this is going to occur if it has not already. The International Atomic Energy Agency (IAEA) reported that there were approximately 140 cases of unauthorized use of nuclear material in 2013 and 160 incidents in 2012 (IAEA, 2014).

Even if law enforcement were somehow miraculously able to shut down the illicit trade in nuclear material there remains the possibility that a rogue state in cooperation with a terrorist group or groups, could unleash such a device. In fact, it may not even be necessary to obtain nuclear material to detonate it. A drone attack upon a facility might set off a reaction that could have disastrous results. The dissemination of technology, especially weapons technology, is not making the world safer. It just leads to weapons in the hands of the worst of humanity with the potential to kill anyone and everyone. Technology has led to a never-ending arms race and a democratization of violence.

In case the reader is not sufficiently alarmed, let one more factor be introduced. The greatest difficulty in terms of combating terrorism is the so-called "lone wolf." This is the individual who acts without connection to any other person or group. Such individuals tend to be able to evade the detection of law enforcement and other security officials tracking terrorists because they do not interact with others.

While the level of knowledge needed is relatively high, and as such somewhat protective, there clearly are people with the ability to manufacture and weaponize biological and chemical weapons. Previously discussed is the evidence that hundreds of such attacks have been carried out by terrorists and criminals as regards biological attacks. Nuclear attacks, due to the complexity of the devices, would appear to be beyond the abilities of a single individual. This assumption is, like many surrounding technologies, not completely grounded in reality and is likely erroneous.

In 1976 a 20-year-old sophomore at Princeton University named John Aristotle Phillips submitted as his class project the design plans for a nuclear bomb (Vollers, 1980). It should be noted that this occurred pre–Internet and he obtained the plans legally by conducting research at the campus library.

11

Evolutionary Dead End
Wrong Turn?

In the end you will die of convenience.—Voltaire

The archeological record reveals that in the history of the planet there have been five periods of mass extinction during which the majority of species in existence were wiped out. Most are familiar with the mass extinction that wiped out the dinosaurs and roughly three-quarters of all other life around 66 million years ago (Borenstein, 2014). In addition to the dinosaurs, four other mass extinctions have occurred as defined by 50 percent of the species dying. Many scientists believe that earth is on the brink, if not already in, the throes of a sixth extinction. Elizabeth Kolbert, in *The Sixth Extinction* (2014), posits that the planet is already past the point of no return. Kolbert places the blame for the current die-off clearly on industrial production and technological advance. The burning of fossil fuels and the accompanying change in climate are leaving numerous species facing certain extinction (Kolbert, 2014). The World Wildlife Fund found numbers even more pessimistic, suggesting that 60 percent of all species have died in the last 50 years (WWF, 2018).

In a study examining the rate of extinction, animal and plant species are becoming extinct at a rate of 100 to 1000 times that before the advent of humans (Pimm et al., 2014). The culprit, according to the authors of the study, is habitat loss. The primary researchers, biologists Stuart Pimm and Clinton Jenkins, place the blame for much of this squarely on human activity. The IPBES (2019) report placed the number of species at imminent risk of extinction at one million.

The accelerating rates of extinction around the world should serve as a clarion call for humanity. Natural evolution is being subverted by a cataclysmic event. This event, the destruction of the natural world, is happening in slow motion from the perspective of a human life. In geological time, this extinction is moving at warp speed.

A key to understanding evolution is to the ability to examine change across vast stretches of almost inconceivable time. From a strictly Darwinian perspective evolution is a slow and gradual process that utilizes natural selection as the means through which one species, or even a trait or set of traits, gain advantage over another. This process is not directed, nor is it planned, as conditions that favor one species over another are difficult to predict and, as oft noted, arbitrary and capricious.

Orthodox Darwinism, also known as uniformitarianism, focuses primarily on the struggle for survival with an emphasis on reproductive success and adaptivity (Darwin, 1859). However, extreme events, such as asteroids crashing into the planet, alter everything. As Kolbert eloquently states, "In times of extreme stress, the whole concept of fitness … loses its meaning. How could a creature be adapted, either well or ill, for conditions it has never encountered in its history?" (Kolbert, 2014).

Unlike the previous extinctions, caused by natural phenomenon, the current and on-going extinction is caused primarily by the actions of one species, *Homo sapiens*. From the beginnings of the Industrial Revolution to the present era of the Digital Revolution, the "progress" of one species is causing the destruction of all species, including its own.

It is strange that the species causing all the destruction is the one that has the ability to reflect upon its actions, to reason, and to understand cause and effect. Unfortunately, this species is also guided by factors other than its ultimate survival and appears unable to take a long-range perspective on its actions.

Examination of the timeline of human evolution reveals that the technological revolution has occurred in no more than the metaphorical blink of an eye. Starting with *Ardipithecus ramidus* at around three to four million years ago, the evolution of humans from bent-over, four-limbed walking to upright, two legs, and talking took something like 90 percent of that journey. If we further examine humanity since our evolution into what is undoubtedly *Homo sapiens*, 90 percent or more of that time was lived as hunter gatherers. It is only in the last 1 percent or less of human existence has there been a significant level of technology—i.e., since the start of the Industrial Revolution—that the evolved physical limits and the culture, both of work and life, began to be altered by machines. Only in the last 10 percent or so of the machine age has the digital age manifested. Its ramifications upon human evolution have thus far barely been considered.

The pace of human evolution continues to be gradual. The pace of innovation and technological development has and is accelerating. This conflict is leading to a necessary alteration in the direction of evolution with unpredictable results.

At the same time the cumulative knowledge of the world is being collected and codified there exists the seemingly contradictory fact that much of the knowledge of the past is being replaced and forgotten. In multiple domains, knowledge that has existed for millennia is being lost. Some things necessarily have a hands-on component. It is one thing to view a web video on fishing; it is an entirely different thing to actually be able to catch fish. A further example concerning fishing serves to illustrate this effect.

In the South Pacific, many of the people of Malaysia supported themselves as subsistence fishermen. The Malaysian fishing industry thrived under these conditions. Japanese fishing conglomerates then entered the picture. Using massive ships that efficiently trawled the seas, these ships directly threatened and competed with the small-scale native fishermen. Eventually the small operators could no longer compete and became employees of the large corporations. After a few decades the area was over-fished and the companies that had taken over the industry left for new areas to pillage. In the meanwhile, the fishermen had sold their boats *and* failed to teach their children the necessary survival skills needed to fish. As knowledge begins to be collected and aggregated, the details begin to be lost. Many of the skills that ancient humans took for granted are being lost forever.

For millennia humans lived as hunter-gatherers and made their homes outdoors, in nature. Today most people would be unable to live outdoors and support themselves through the hunter-gatherer lifestyle no matter what the circumstances that forced them to do so. Several video productions, including *I Caveman* (Heitman, 2011), with Morgan Spurlock, and the survivalist series *Naked and Afraid* (2013–present) demonstrate this in an unintentionally humorous way. In each of these productions modern humans attempt to live lifestyles akin to that of hunter-gatherers. Mostly these efforts are noteworthy for how poorly equipped the people, even the survivalists, are for living in nature. The primary lesson from these efforts is that the necessary skills for successful life outdoors have simply been lost. As humans become more and more divorced from their evolutionary past and path, a risky bet is being made on which skills are most important. At present, humans are experiencing a "de-skilling" of previously acquired abilities. The effects of technology on the future evolution of Homo sapiens are yet to be seen.

One factor that appears consistently related to a species' ability to adapt and survive is diversity. Whether this diversity is endogenous (e.g., genetic) or exogenous (e.g., food sources) the existence of a range of options appears beneficial. From this perspective and multiple other metrics, the modern world would seem to be on a path towards disaster. For example, the number of cultures, as represented by the number of languages spoken, is declining at a precipitous rate. In 2009 there were an estimated 7,000 languages spoken in the world (Ethnologue, 2009). This number is predicted to decline by nearly half with 3,000 languages expected to die out over the next century (Language Matters, 2015). As languages disappear, a view of the world, a unique perspective on knowledge, also dies. Other languages are also being modified as certain words of the techno-cyber culture have begun to infiltrate them in much the same manner as Latin spread under the Roman Empire.

Languages are not the only phenomenon to demonstrate a decrease in diversity. Even food has become uniform and regularized in terms of shape and color. While the global food chain gives the appearance of greater choice the reality is that for numerous species from bananas to tomatoes subspecies have become imperiled and have disappeared because of the demands of globalization. Bananas have declined from over 400 species to one (Keoppel, 2008; 2011). Tomatoes have lost flavor because of efforts to enhance shelf life, appearance and transportation ease (Warner, 2013). The turn to monoculture as the dominant means of food production is not only affecting food but the environment as well. As noted in the 2019 IPCC report, monoculture tends to weaken and even destroy the wildlife and the land on which it is practiced, leading to even further degradation.

Food, and increasingly the production of it, is becoming specialized in such a manner that fewer and fewer individuals are capable of harvesting and producing food. While multiple new methods have been developed to process food and produce "food-like" substances that provide calories and occasionally nutrition, the ability to plant crops and live a subsistence lifestyle is disappearing. Food is but one manifestation of this reality.

In the past, food scarcity was a reality and that is why we now have obesity. Humans evolved with a body geared towards seeking food and a brain with a poor "full" sensor. Even with a relative abundance of available food, the procurement of enough to sustain an individual was a daily chore. The collection and distribution of food among social animals accompanied the development of culture among *Homo sapiens* in such a manner that many societies appear to center around food. Communal meals appear to be one of the few cultural universals.

Millennia of evolution are now being short circuited with the wide availability of easy to acquire food. It is not the wide availability of food that is leading to epidemics of obesity, diabetes and autoimmune diseases, so much as what we are doing to the food. Specifically, more and more evidence is accumulating that the processing of food is causing health problems and the convenient, ready availability feature loved by consumers is the very thing that is killing them slowly with a knife and fork.

You Are What You Eat

The meme of the pill as food is a common one throughout the history of the science fiction genre. Even in the space themed cartoon of the 1960s, *The Jetsons*, food was in the form of pills. Forget the enjoyment of eating in and of itself, the idea that a pill could substitute for food was uncritically presented as an example of progress. Supplements aside, the pill as food never really took off in the mind of the public. But making food preparation easy and instantaneous did. Canned food and microwave meals appeared in the 1950s and 1960s—representatives of the future. The near simultaneous creation of fast food chains changed the way Americans ate in less than a generation. Prior to World War II almost all meals consumed by the average person were made in the home, more or less from scratch. By the end of the Korean War the availability of easy and quick food was taken for granted. Food has moved from being the cross-cultural universal to which every family unit devoted a great deal of time in procurement and preparation to a commoditized transaction in which the food is prepared out of sight of the consumer.

A text examining modern eating habits, *Pandora's Lunchbox* (Warner, 2013), posits that the very nature of our instantaneous food culture, is leading to illness. Focusing on processed food, the author notes that the American diet is 70 percent processed foods and that life expectancy is 37th in world ranking. Numerous studies are presented that demonstrate the association of processed foods to health problems and the association of a healthy, plant-based, non–processed foods diet to longevity. The food industry has been remarkable in hiding the deleterious effects of the modern, instantaneous diet. Economic interests far outweigh, and far outspend, health concerns.

According to Warner (2013), the problem lies in the fact that we have a "Stone Age Body." Like our brains, our bodies have been relatively unchanged for many thousands of years. The introduction of processed grains, preservatives, salt, refined sugar, etc., into our diets is having a deleterious effect upon health and have also been linked to increases in certain diseases such as celiac disease and multiple chemical sensitivity.

Astonishingly, some individuals appear to be taking this process in the opposite direction. A food-like product called "Soylent" has been developed that serves as the ultimate representation of food as technology concept. Containing what are deemed to be the essential nutrients, vitamins and calories necessary to function, this product serves to reduce food to a functional process that has nothing to do with sharing meals, community or even the enjoyment of eating. It is food made to eat on the go, food for those with no time, used only to meet the necessities of life.

The New World

When explorers first arrived in the New World from Europe they quickly came to dominate the native populations primarily through superior weapons and an ability to disregard the humanity of those whom they subjugated. The difference in technological development greatly aided colonialism and the extraction of resources from the newly "discovered" land (where 100 million people lived).

The difference in technological development continues among the people of the world even today with a broad range of acceptance and adoption of various innovations. It should be noted that even with ubiquitous technology not every invention is universally accepted. While adoption of new technology is happening at a faster rate it is not an absolute that every individual will adopt every technology even within a culture.

This differential adoption of technology brings attention back to a well-established and previously discussed concept in the fields of sociology and anthropology known as cultural lag (Ogburn, 1964). This concept seeks to explain the cultural disruption that can ensue when certain elements of the culture, such as technology, change at a faster pace than other elements of the culture. Beyond being a simple process that affects everyone equally, cultural lag can have differential effects on disparate elements of society, including areas seemingly unrelated to each other, and can extend to matters outside of technology. For example, multiple movements in music, such as jazz and rock and roll, caused cultural upheaval and split people along newly emerging cultural lines.

In the realm of technology, most innovation affects relatively small sub-cultures and is therefore not disruptive on a large scale. Examples of this type of cultural change relate to tools and innovations in certain business applications that may change how a certain product is produced or a certain production process is completed. In this case, only those directly involved in the product may even notice the change, but it might be completely disruptive in that industry. An example of this might be a new process for manufacturing that gives one company an insurmountable advantage over its competitors.

When it comes to the digital age and computers, the effects of cultural lag are enormous. Partially because of the ubiquitous nature of computers in our lives, the cultural disruption and cultural divide(s) are emerging along multiple lines. It is a common observation that older people are less comfortable with new technology than younger people. However, the digital age is splitting people along many other lines in addition to age. Among these are wealth, education, ethnicity, gender and, likely, personality.

In the digital age, being connected is viewed as an advantage and many of the people without connectivity are so because of financial disadvantage not because of a thoughtful rejection of technology. Yet, there are those who purposely and deliberately reject certain technologies. Certain cultures, some of which might be unaware of the rest of the world, some of which seem to have made a collective decision not to participate in the modern world, continue to live without that which most people see as indispensable.

Suppose that the concept of cultural lag goes deeper than disadvantage? Might it be that differential development might be akin to diversity among different sub-species of animals? The suggestion herein is that cultural lag may have its roots in evolution. What is being posited is that the manifestation of different cultures embracing technology to different degrees might be a protective mechanism that serves as an evolutionary safeguard against extinction.

There once was a primate, *Gigantopithicus*, which essentially failed because it was too large. Humans, on the other hand, might be on the road to failure as a species as a result of becoming too reliant on machines and forgetting how to live otherwise. What is being suggested, what is being feared, is that humanity has turned off the road to evolutionary survival and has instead, through cultural dominance and religious-like fervor that demands an unquestioning acceptance of all manner of technology, embraced the agent of its own demise. The road to extinction appears to be the road to utopia until unexpected events occur.

In an evolutionary sense, the digital divide is about more than the possession of advanced technology. The true divide may be the one that is not seen. It could be that the real divide is between survival and extinction. Perhaps groups not enthralled by technology are creating different pathways of evolution for humanity. Futurists and extropians will pull one branch of humanity closer to machines while uncontaminated tribal peoples may continue a more naturalistic/organic evolution. Random events of the future will determine who survives. If a solar flare kills everything electronic, then those who have avoided technology will have an advantage. If all of nature is destroyed, including the habitat of those not dependent on technology, then the technophiles will have a momentary advantage. At present the technophiles seem to be trying to stack the deck by killing the natural world without understanding that the technological world is dependent upon the natural world. However, as changes in climate are bringing to attention, nature bats last.

Preventing everyone from going in the same direction might ultimately be what saves the species. It is held that at one point human population may have dropped to as low as a 2000–3000 individuals somewhere around 50,000 to 75,000 years ago (Hawks, Hunley, Lee, & Wolpoff, 2000).

For reasons not completely understood, nearly everyone alive was wiped out except for a small population clustered in what is now northeast Africa. Why they survived while everyone else died is debated but the fact remains that were it not for this group, humanity would have become extinct at that time.

Consider what happens in the event of something such as massive solar flares that disrupt or destroy all communication and electronic devices. No doubt there would be huge numbers of deaths and social disruption. Those who will not notice or who will go on as before will be the so-called primitive people who are now out of the loop.

There continue to be various isolated groups of people around the globe who utilize very little technology. Some groups, such as the Amish, consciously avoid certain technologies and draw a line at a certain level of technological advance. In other cases, such as a few remaining Stone Age tribes, technological development has remained limited. It may be however, that there is much to learn regarding our dependence on technology when we more closely examine these groups. Maybe if or when humanity bombs itself back to the Stone Age the few remaining Stone Age tribes will become the models of how to live.

In the Andaman and Nicobar Islands in the Indian Ocean, between India and Indonesia, there are several islands inhabited by Stone Age tribes. Primarily because of rumors that they are headhunters, the modern world has left them relatively untouched. When the tsunami occurred in Indonesia in 2004, following a massive earthquake measuring 9.1–9.3 on the Richter Scale (U.S. Geological Survey), these islands were right in its path. This tsunami killed an estimated one-quarter to one-half million people and devastated a large portion of the island of Sumatra (BBC, 2005). In contrast, the inhabitants of these islands, the alleged primitives, the Stone Agers, emerged relatively unscathed. When the Indian government sent forces out in helicopters to check on those living on North Sentinel Island on what they thought would be a devastated population caught unaware, they were instead greeted by the islanders shooting arrows and throwing spears at them, preventing them from landing (*Times of India*, 2005; National Geographic Society, 2005). Satellite images later confirmed that the inhabitants of the island recognized that a tsunami was imminent and had fled to high ground long before they were imperiled (European Space Agency, 2005). It seems they were in touch enough with nature to recognize what sophisticated monitoring equipment did not.

It has been noted that there are tribes in the rainforests of South America that have not yet been contacted by "modern" civilization and appear to be actively avoiding such contact by moving and camouflaging their camps (Hammer, 2013). There are also other groups that are aware of the modern world and whose tribes have contacted the outer world but are now trying to avoid further contact and appear to be eschewing certain forms, but not all, of technology.

There is at least one precedent for a group developing a degree of technology and then turning its back on it. Australian aborigines who lived on the island of Tasmania completely eschewed all types of technology at one point in their history going so far as to throw away fishing hooks (Lawlor, 1991). Philosophically, they came to believe that technology took them out of touch with nature and imperiled their existence. The traditional aborigine belief system appears oriented towards an understanding of sustainability and the need to follow the designs of nature. This group consciously decided that technological development was at odds with their beliefs and followed their philosophy to its logical conclusion.

Efforts to wire the planet or provide solar computers to everyone are well intentioned but also dismissive of ideas of cultural determination. These efforts also ignore the effects of cultural disruption that might result from bringing the digital revolution to pre–Bronze Age cultures. The idea that these groups may not desire the trappings of modernity is barely considered. Again, similar to Conquistadors and Crusaders, the current dominant techno-culture never stops to consider the implications of the imposition of a worldwide view of better living through radical connectivity and advanced technology. At some level it is almost as if indigenous people are being told, "Here, have a computer as payment for the destruction of your home." This process of cultural assimilation is firmly rooted in a philosophical position that sees apparent success as the arbiter of good and bad.

Putting all of the metaphorical eggs of survival in the basket of technology is similar to investing all of one's money in one stock. This would not be viewed as a sound investment strategy. Survival is often a matter of adaption and happenstance. Intelligence appears to be a great evolutionary advantage, but for sharks and insects and countless other species intelligence seems to play a negligible part in their success. Environmental match and fitness appear far more related to survival. The view that technological advance is a symbol of progress is philosophically bankrupt. Ideas that technology will save humanity are more rooted in cultural myth and the ethnocentrism of the dominant techno-society than in an evaluation of outcomes. Eight million people a year dying as a direct result of pollution would hardly be positive evidence of the proposition of superior living through technology. Yet despite the many pitfalls of a technological existence, the cultural imposition continues to occur. Obvious cultural relativism is ignored by the techno-society and those not fully immersed are viewed as "less than" and subjected to terms such as "primitive" and "backward."

Perhaps these individuals denigrated as primitive have not bought into the myth of progress. Nature, as noted, employs a simple win/stay, lose/shift, strategy. A case can be made that those who still exist in so-called primitive societies have been the evolutionary winners and have had no need to change or innovate. Following this train of logic, modern netizens, those most technologically advanced, may be demonstrating themselves to be evolutionary losers who need to keep shifting strategies, inventing new technologies, in order to improve outcomes. Those who appear to exist "out of time" are living much closer to the way humans have lived throughout history and prehistory than the involuntary cyborg of today.

Consider a joke about a Kung! Bushman given a laptop. Presented to him personally by Mark Zuckerberg and Michael Dell as part of the Laptops to the World Project, the man initially appeared puzzled by the gift. A month later a reporter tracked him down and found he was effusive in his thanks. "It was a very good gift," he told the reporter. "I was able to use all the parts of it for very useful things." He proudly displayed a snare he had made with the wires and a frame for tanning animal hides constructed from elements of the casing. The hard-plastic edges made great scraping devices and the screen was used to magnify the sun's rays for cooking. "Too bad I don't have electricity," he added, "who knows what I could have done with this machine."

That is adaptability. That is intelligence. That is survival.

Consciously or not, there does seem to be some awareness of the need to remain in contact with our more "human" roots. Events like Burning Man or Rainbow Gatherings, while often derided by mainstream culture as "hippie," or "weirdo," events, at core these appear to be about community and family and hearken directly to tribalism. The simplicity movement of the transcendentalists, a direct reaction against the Industrial Revolution, as well as the modern mindfulness movement appear to be a reaction against the credo of better living through machines. The "modern primitives" subculture likewise bears direct testament to the need for a less technological existence (Vale & Juno, 1989). Even a few peer reviewed studies can be found that emphasize the importance of experiencing the natural world (Russell et al., 2013; Wolsko & Lindberg, 2013). "Green bathing," spending time in nature, has now become prescriptive.

The Dead Earth subgenre of science fiction might also point to a collective unease regarding the technological omnipotence that pervades modern life. "Dead Earth" refers to a recurring storyline in science fiction where Earth has become uninhabitable for a wide range of reasons including cataclysmic events and environmental destruction. The destruction is often caused by human activity intended to improve life. It is typical for the destruction to be a result of the by-products of the technology that was thought to be lifesaving and beneficial. See *Cat's Cradle* (1963) by Kurt Vonnegut; *The Martian Chronicles* (1962) by Ray Bradbury; and the *Dirt People* (2005) by Ray Bawarchi.

Human arrogance suggests the solution is not to question the whole notion of progress, nor to even hint at technological overreach, but rather to double down. When technology has become a deity, to seek a solution that is non-technological is to become a heretic. One of the more unfortunate implications of the belief in the ability to colonize Mars or beyond is the idea that taking care of this planet is no longer important. The moment humanity decides it can leave Earth it will begin to treat the planet like a garbage can. Tragically, it seems that the decision has already been made by many. As an aside, the whole notion of leaving the planet of origin sounds decidedly counter evolution. It is not the inability of nature to absorb all the pollution that is the problem, but rather the production of the waste in the first place. Waste disposal will also be a problem for people living in bubbles on Mars or Titan. Technological true believers see only solutions. Steven Hawking, to his credit, expressed some reservations regarding space exploration. Specifically, he warned that attracting attention from other civilizations exploring the galaxy may prove disastrous. He noted that explorers on this planet tend to be imperialistic and generalizes that this may be true of other species as well. He also warned that any civilization that could contact Earth would necessarily be more technologically advanced and as such could likely overrun the planet should they so desire (Colfield, 2015).

A more radical, although perhaps entirely logical and reasonable, view of technology is that it is ultimately leading humanity to its doom. Frank Fenner, emeritus professor of microbiology at the University of Canberra, suggests that technological innovation necessarily ends in destruction and obliteration. He predicts that humans will become extinct as a result of environmental destruction and the inability to sustain energy demands (Edwards, 2010). In a more expansive argument grounded in the violence of weapons technology and the ever-increasing race to build faster, larger, more efficient and destructive killing machines, Lord Martin Rees (2003a; 2003b) predicted, with a 50 percent level of confidence, that humanity has only about 95 years left. The Self Destruction or Doomsday Argument hypothesizes that all technologically based cultures eventually reach a point where they blow themselves up (Hart, 1972). This idea posits a timeline that suggests that after the milestone of the nuclear age is reached; all societies wipe themselves out after 100 to 200 years. This idea was initially proposed as a solution to the Fermi Paradox, to explain the lack of detected alien life in the universe. Hart (1972) suggested this type of destruction might have happened numerous times in the history of the universe. So even if all the barriers to space travel are overcome, finding other life forms might be even more difficult than initially thought. It might be that they just are not there anymore.

This is not to imply that the direction of human evolution and behavior is guided by conscious thought. One of the important contributions of the positivists and their intellectual descendants, the behaviorists, is the undeniable fact that a great deal of the behavior in which humans engage is neither thoughtful nor even thought about. Much of what humans do is mechanistic and there is little to no reflection about it. Some functions at the neurological level serve purposes that are difficult to ascertain. Some served functions in our ancient past that no longer have modern analogs. Other behaviors adapt well to different contingencies.

An examination of this idea might explain human fascination with the television, computers and phones. Analogous to ancient humans sitting around the fire, modern humans stare at devices and instead of telling each other stories, the fire (devices) now tells the story. Television is a perfect example of the upside and downside to technological innovation. Television can be, and occasionally is, a source of information. It has been used to educate children, teach languages, and provide a steady stream of news. But mostly it is a source of entertainment, often mindless and insipid, a perfect medium to introduce products to consumers. People watch largely out of habit, looking for something interesting rather than tuning in to watch anything in particular. Computers and phones now serve to provide equal and pointless analogs.

In the past we scanned the environment looking for peril and adventure. Scanning and searching online now serve the same function. Surfing on the computer is like scanning the landscape. Searching is more about the search than the find. One thing leads to another. Search upon search, on and on. For our ancient ancestors this searching served a function, for modern humans, it is nothing more than a waste of time.

Working constantly with our hands to do and make things is a natural behavior for people. Using one's hands texting, guiding a mouse—are all modern analogs. Shelling peas, shucking corn, weaving baskets and other such behaviors from the past served similar functions. They all involved the hands being active but little mental processing was required.

The fact that certain behaviors fill a void in evolved behavior that no longer has an outlet suggests the underlying factors leading to computer/cyber addiction are neurological. Humans need to do certain behaviors at a physical level. It doesn't matter what we do, we just need to do. We don't tan hides or milk cows anymore; we text and surf.

Unfortunately, some evolutionary presets do not translate so well. The cognitive tendency towards suspicion and paranoia towards others underpins a mentality that leads to an arms race to prepare for eternal conflict.

When archaic humans were fighting with stone axes, the result was brutal, but it was also limited both geographically and numerically. Today this paranoia has the potential to lead to complete annihilation.

Humans evolved with a particular set of constraints. These evolutionary givens are sorely taxed in the modern error (misspelling deliberate). There are evolved limits on the ability to process information, the capacity to attend and remember, to catalog people, and the cognitive space needed to carry out activities and complete tasks.

We have a brain that is wired for the lifestyle of a hunter-gatherer. Our consciousness is constantly on alert for novelty within the landscape. Change signifies that the normal state has been altered. Enter a device that can provide a steady stream of novelty. Energy that would have been devoted to scanning the landscape for food or danger is now channeled into bouncing from one website to another. A primary need at a neurological level seems to be met, yet there is no payoff to the action, no food captured, nor predator eluded. The drive state is aroused and maintained, yet there is no final resolution because the goal is no longer specified.

Even in the specific case of sexual release, orgasm does not provide resolution of the behavior. At a basic level this is because masturbation is a substitute for intercourse. Searching for pornography online serves an evolutionary drive of mate evaluation and selection. In most cases, Internet dating sites aside, online pornography provides unattainable choices. The outcome is the already documented pattern of regular consumers of pornography becoming more addicted to the search for the images than to the actual viewing of them. Zimbardo (2011) has proposed "Arousal Syndrome" to identify this behavior. Before porn, men had to go out and attempt to meet actual women. The use of gender terms herein is deliberate. That men are more prone to porn addiction is likely a result of biological presets related to mate selection. The cyber world can essentially short-circuit the biological imperative. Drive states are met through a poor analog, in this case a digital analog, to the real experience.

It is clear this type of "arousal" could be generalized and extended to a large degree of online behaviors that clearly have problematic components. Looking for pictures of cute kittens could theoretically be as addictive as pornography. However, judging by sheer volume, porn addiction would seem to be a major issue. The largest site gets nearly nine million page views per day and several other sites have huge volumes of traffic as well (The Chive, 2017). According to some sources, up to 37 percent of sites are dedicated to the pornography business (Optonet Website, n.d.). Even more modest estimates range up to 30 percent (Anthony, 2012).

Sex, food, and safety. The World Wide Web can provide illusions of all the above. The being that evolved over a couple of million years is now propelled into a strange universe where only poor substitutes exist to fulfill subconscious needs. Instantaneous gratification serves to validate a virtual existence. When the space program began monkeys were strapped into capsules and fired into space. This phenomenon continues to this day, but the monkey has now decided it is more than a naked ape. This time it thinks it is becoming a transcendent being.

Humans have now entered the machine. In some cases, such as with Big Data, humans may even be the machine, little more than nodes on a network. What effect will this development have on the path of evolution? Will the reliance on technology prove beneficial or harmful? Is this the path forward or a dead end? This raises the question of whether the technological context in which we are living is altering the way in which we are evolving. The answer is almost certainly that it is altering our future development but how? Are we becoming ever more powerful or are we becoming ghosts, faint traces with no substance, lost in the machine?

Humans have already developed the technology to alter genes and therefore the direction of evolution. In England gene-splicing technology that allows for the creation of babies with three parents was legalized in 2015. Two out of every 100 births in the U.S. are the product of reproductive technology (Stephen, Chandra & King, 2016). In some instances, employers, e.g., Apple and Facebook, are paying for female workers to freeze their eggs to prolong their work careers while extending reproductive viability (Tran, 2014). It is already possible to buy the eggs of supermodels on the Internet and to get the sperm of Nobel Prize winners. Human engineering is not just a future possibility it is a present reality. This reality must be confronted. Well intentioned practitioners and researchers can create scenarios where only the positives of this manipulation will be emphasized. It is the height of human hubris to believe that the manipulation of nature is something without risk. In many nations the law is simply silent on such issues. This essentially means that there is no law and all that limits those without ethics is money.

A recently developed method for genetic manipulation is a process known as CRISPR (Clustered, Regularly Interspaced, Short Palindromic Repeats), whereby certain gene sequences can be manipulated for specific purposes. The ethical implications raised by such technology are enormous; this process would make it easier to engineer and weaponize biological organisms. The mere ability to do this has apparently blinded researchers to the implications. Esvelt, the lead researcher in the studies conducted at MIT, hailed the process as "accelerating evolution." Alarmingly, Britain's Human Fertilization and Embryology Authority approved gene-editing experiments despite significant outcry from ethicists (Cheng, 2016).

The major ethical issue associated with CRISPR (Clustered Regularly Interspaced Short Palindromic Repeats) is that it allows humans to alter their own evolution. Not only are the genes in the person being altered but so are the ones they pass along. The medical community is poorly equipped to look at this issue. Their concern is with alleviating illness and pain. This bias is all too clear in a report from the National Academies of Science and Medicine that generally questioned the use of gene-editing for inherited diseases but then stated support for research into gene-editing of inherited diseases "only" in the case of serious disabilities and diseases. One can readily see that this loophole will be available for every disease.

The lack of control over such technology and the likelihood of unethical behavior has already been realized. He Jiankui, a geneticist in China, has used CRISPR to immunize twins against HIV (Bi, 2019). He did this despite government prohibitions against the practice and subsequent nearly worldwide condemnation. More remarkable than what he did is the reaction of surprise. David Liu, who developed a variant of the gene-editing tool, stated, "It's an appalling example of what not to do about a promising technology that has great potential to benefit society. I hope it never happens again." Is such naiveté about human behavior possible? To demonstrate how far out of bounds gene technology has already progressed is a report from China concerning Bing Su of the KunMing Institute of Zoology who genetically altered the genes of monkeys by injecting a virus that affected brain development. Despite the ethical outcry, researchers brushed away the concerns (Burrell, 2020).

The strange thing about evolution is that success is defined only after the fact. Technological development might prove an assist to quantum leaps in consciousness or it might create reliance and dependence in exactly the manner that leads to our extinction. Humans might figure out nuclear power and produce limitless energy for the whole world or instead might blow up the whole world by losing control of the nuclear genie. Given the history of humanity it is hard to be optimistic.

Reliance on machines seems to be making people forget how to do things they could previously do. A prime example would be the use of GPS/Smart phones to navigate. Because of these devices most people no longer carry maps in their cars. As a result, when the system is down, there is no way to proceed. This effect is compounded because, as a general rule, areas of sparse coverage are also the most difficult to navigate without maps.

Another related example has to do with the forced availability, or non-availability, of pay phones. In the not too distant past if one needed to use a phone in an emergency, one could almost always be found. Now, with the widespread use of cell phones, pay phones have become nonexistent.

Try to remember the last time you saw one. What do you do in an emergency if your phone is damaged?

Is it possible to become so reliant on something, so over reliant, that it becomes a dead-end trap? Using devices when you don't need them makes you weak. It is analogous to a person using a wheelchair when able to walk. The muscles will slowly atrophy and eventually the person will be unable to walk.

The history of extinction is replete with species that became so overly specialized that rapidly changing conditions wiped them out. Others, metaphorically, painted themselves into a corner and had nowhere to go when they ran out of room.

Is technology destined to become to humans what bamboo is to pandas, what eucalyptus is to koalas? At least some evidence suggests humans are already headed in that direction. An effort to use and develop mechanical bee-like drones to serve as pollinators has begun. The idea is that because bees are dying out, building small machines to take their place is the proper course of action. Problems so far include the machines killing the plant and crushing the pollen (Khan, 2017; Chechetka, Yu, Tange & Miyako, 2017).

Every science fiction writer worth his or her Klingon salt has considered the possibility of the machines taking over. In SF it's always with a bang. But what if it's with a whimper? Could a slow insidious process make us so dependent on machines that we forget how to do anything (like navigate)—and then could we lose the ability to repair the machines?

In an episode of the classic sci-fi series *The Twilight Zone*, humans welcomed what appeared to be benevolent advanced aliens to Earth. A book with the title "To Serve Man" is found with the aliens. Humans begin translation of the book anticipating all the advances the superior alien technology will bring. Just as vast numbers of humans have boarded ships to go to the aliens' home planet the rest of the book is translated. As the ships blast off a frantic voice cries out, "It's a cookbook."

Could humanity now be in such a moment as regards technological advance? Is humanity racing headlong into a seemingly miraculous future that may ultimately prove to be the element of our demise? Are the devices that are intended to provide us with wealth, leisure and knowledge leading us down the road to ruin? As we poison our planet and subvert our own evolution, the belief in a magical future where technology solves every problem allows us to avoid the hard work. Rather than changing our behavior, it is easier to ignore the magnitude and reality of problems like climate change and mass extinction by trusting that future technologies will make everything okay.

Betting the future on unproven and undeveloped technology is an act of misplaced faith. Like religious zealots who run into a hail of bullets because they believe god will protect them, technophiles exhibit similar irrational beliefs. Hailing technology as the savior of the planet and the instrument of every solution is misplaced devotion. Let us not let the worship of technology become a martyrdom mission.

Rather than attempting to become one with the machine, perhaps we should ask ourselves what is wrong with being a human. Sometimes the solution is less, not more.

Bibliography

Adams, L. (2017). Dementia: Could heavy traffic drive dementia risk? *Nature Reviews Neurology*, *13*(3), 128.

Adler, J., and J. Hersey. (2015, May). Mind meld. *Smithsonian Magazine*, *46*(2), 44–51.

Agarwal, A., and T.M. Said. (2003). Role of sperm chromatin abnormalities and DNA damage in male infertility. *Human Reproduction Update*, *9*(4), 331–345.

Agarwal, A., F. Deepinder, R.K. Sharma, G. Ranga, and J. Li. (2008). Effect of cell phone usage on semen analysis in men attending infertility clinic: An observational study. *Fertility and Sterility*, *89*(1), 124–128.

Ahlborg, U.G., A. Brouwer, M.A. Fingerhut, J.L. Jacobson, S.W. Jacobson, S.W. Kennedy, ... and S.H. Safe. (1992). Impact of polychlorinated dibenzo-p-dioxins, dibenzofurans, and biphenyls on human and environmental health, with special emphasis on application of the toxic equivalency factor concept. *European Journal of Pharmacology: Environmental Toxicology and Pharmacology*, *228*(4), 179–109.

Akpan, N. (2015, December 10). These test-tube puppies, the world's first, may save endangered wolves. *PBS NewsHour*. Retrieved from https://www.pbs.org/newshour/science/worlds-first-test-tube-dogs-in-vitro-fertilization-endangered-wolves.

Akpan, N. (2016, February 11). How a litter of puppies could help save endangered animals. *PBS NewsHour*.

Aksglaede, L., A. Juul, H. Leffers, N.E. Skakkebaek and A.M. Andersson. (2006). The sensitivity of the child to sex steroids: Possible impact of exogenous estrogens. *Human Reproduction Update*, *12*(4), 341–349.

Alivisatos, A.P., M. Chun, G.M. Church, R.J. Greenspan, M.L. Roukes and R. Yuste. (2012). The brain activity map project and the challenge of functional connectomics. *Neuron*, *74*(6), 970–974. doi:10.1016/j.neuron.2012.06.006.

Alpers, P., A. Rossetti, D. Salinas and M. Wilson. (2015). *Rate of civilian firearm possession per 100 population. United States—gun facts, figures and the law*. Sydney, Australia: Sydney School of Public Health, The University of Sydney.

American Automobile Association (AAA) (n.d.). What does it cost to own and operate a car. Retrieved from https://www.aaa.com/autorepair/articles/what-does-it-cost-to-own-and-operate-a-car.

American Civil Liberties Union (ACLU) (n.d.). Retrieved from https://www.aclu.org/.

American Civil Liberties Union (ACLU). (2015). *FISA. STAND 2015*. Washington, D.C: ACLU.

The American Heritage Dictionary, Fourth Edition (2002). Boston: Houghton Mifflin Company.

American Meteorological Society. (2014). Explaining extreme events of 2014 from a climate perspective." *Bulletin of the American Meteorological Society*, *96*(12). https://journals.ametsoc.org/doi/pdf/10.1175/BAMS-ExplainingExtremeEvents2014.1.

American Meteorological Society. (2014, September). *Bulletin of the American Meteorological Society*, *95*(9).

American Psychiatric Association (APA). (2013). *Diagnostic and statistical manual of mental disorders- 5th edition* (DSM-5). Washington, D.C.: American Psychiatric Publishing.

American Psychiatric Association (APA). (2015). Depersonalization/Derealization disorder. In *Diagnostic and Statistical Manual of Mental Disorders, 5th edition*. Washington, D.C.: American Psychiatric Association.

American Psychological Association (APA). (2019). Code of Ethics. Retrieved from https:// www.apa.org/.

Anderson, S. E., and A. Must. (2005). Interpreting the continued decline in the average age at menarche: Results from two nationally representative surveys of US girls studied 10 years apart. *The Journal of Pediatrics, 147*(6), 753–760.

Apuzzo, M. (2016, November 16). Some U.S. phones send text messages to China. *Atlanta Journal-Constitution*, p. A2.

Arbuckle, T. E., R. Hauser, S.H. Swan, C.S. Mao, M.P. Longnecker, K.M. Main, ... and C. Till. (2008). Meeting report: Measuring endocrine-sensitive endpoints within the first years of life. *Environmental Health Perspectives, 116*(7), 948–951.

Ariel, B., W.A. Farrar and A. Sutherland. (2015). The effect of police body-worn cameras on use of force and citizens' complaints against the police: A randomized controlled trial. *Journal of Quantitative Criminology, 31*(3), 509–535.

Armstrong, L., J.G. Phillips and L.L. Saling. (2000). Potential determinants of heavier internet usage. *International Journal of Human-Computer Studies, 53*(4), 537–550.

Aronson, G. (2013, June 3). Supreme court okays warrantless DNA sampling. *Mother Jones*.

Asimov, I. (1942). *Runaround*. New York: Street and Smith.

Asimov, I. (1950). *I, robot*. New York: Doubleday.

Aspell, P.S. (Executive producer). (2015). *NOVA* [Television series], Cyberwar threat [Series episode]. United States: WGBH Boston.

Aspell, P.S. (Executive producer), and Dart, K (Director). (2014). *NOVA* [Television series], Rise of the hackers [Series episode]. United States: WGBH Boston.

Associated Press. (2015, February 26). "Bionic reconstruction gives men first bionic hands controlled by mind." *Atlanta Journal Constitution*, A8.

Associated Press. (2015, July 2). VW robot connected to contractor's death. *Atlanta Journal-Constitution*, p. A10.

Associated Press. (2016, March 26). Drone delivers item to residential area. *Atlanta Journal-Constitution*, p. A10.

Astill, K., and M. Griffith. (2014) *Clean up your computer: working conditions in the electronics sector*. (A Catholic Agency for Overseas Development). Retrieved from https://goodelectronics.org/wp-content/uploads/sites/3/2004/01/Clean-up-your-computer.pdf.

Atlanta Journal-Constitution. (2019, March 19). Plastic in whale. *Atlanta Journal-Constitution*, p. A4.

Atlanta Journal-Constitution. (2019, April 2). Agency sounds alarm after 48 lbs. of plastic found in dead whale. *Atlanta Journal-Constitution*, p. A4.

Azucar, D., D. Marengo and M. Settanni. (2018). Predicting the Big 5 personality traits from digital footprints on social media: A meta-analysis. *Personality and Individual Differences, 124*, 150–159.

Babwin, D. (2016, July 17). Experts: Use of robot to kill suspect opens door for others. *Atlanta Journal-Constitution*, p. A6.

Bach, J.F. (2018). The hygiene hypothesis in autoimmunity: The role of pathogens and commensals. *Nature Reviews Immunology, 18*(2), 105.

Bacon, F. (1994). *Novum organum*. (P. Urbach and J. Gibson, Trans.). Open Court. (Original work published 1620).

Baker, B. (2013, December 27). *70+ USS Ronald Reagan crew members, half suffering from cancer, to sue TEPCO for Fukushima radiation poisoning*. ECOWatch. Retrieved on November 19, 2015 from http://ecowatch.com/2013/12/27/ronald-reagan-cancer-sue-tepco-fukushima-radiation/.

Baker, L.R. (2013). Technology and the future of persons. *The Monist, 96*(l), 37–53.

Baldwin, A., S. Corsi, P. Lenaker, M. Lutz and S. Mason. (2015). *Beyond microbeads: Microplastics in Great Lakes tributaries*. Reston, VA: USGS: Great Lakes Restoration.

Barabasi, A. (2003). *Linked: How everything is connected to everything else and what it means for business, science, and everyday life*. New York: Plume.

Barbier, E.B., S.D. Hacker, C. Kennedy, E.W. Koch, A.C. Stier and B.R. Silliman. (2011). The value of estuarine and coastal ecosystem services. *Ecological Monographs, 81*(2), 169–193.

Bare, R. (Director). (1962). *The twilight zone* [Television series], To serve man [Series episode]. United States: Cayuga Productions.

Bar-On, Y.O., R. Phillips and R. Milo. (2018, May 21). The biomass distribution on earth. *PNAS, 115*(25), 6506–6511. doi.org/10.1073/pnas.1711842115.

Bawarchi, R. (2007). *The dirt people.* Asheville, NC: Blue Throat Press.

Bawarchi, R. (2012). *Conflict ahead for India, Myanmar: Drowning Bangladesh.* India America Today. Retrieved from https://www.indiaamericatoday.com/article/conflict-ahead-for-india-myanmar-drowning-bangladesh/.

Bean, D. (2013, March 13). Facebook "likes" used to predict personal information. *ABC News.* Retrieved on August 12, 2014 from http://abcnews.go.com/blogs/technology/2013/03/facebook-likes-used-to-predict-personalinformation/.

Beans, L. (2013, July 1). *Genetically engineered crops trigger cycle of superweeds and toxic pesticides.* Retrieved from https://www.ecowatch.com/genetically-engineered-crops-trigger-cycle-of-superweeds-and-toxic-pes-1881771901.html.

Becerra, T.A., M. Wilhelm, J. Olsen, M. Cockburn and B. Ritz. (2013). Ambient air pollution and autism in Los Angeles County, California. *Environmental Health Perspectives, 121*(3), 380.

Bee Informed Partnership. (n.d.). Retrieved from https://beeinformed.org/.

Begley, S. (2013, March 5). "Nightmare bacteria," shrugging off antibiotics, on rise in U.S. *Reuters.* Retrieved from http://www.reuters.com/aricle/2013/03/05/us-bacteria-idUS BRE924.

Benneton, L. (2008, June 11). *Moore's Law and effect on the auto industry.* Retrieved from http://luigibenetton.com/2008/06/moores-law-and-effect-on-the-auto-industry/.

Bennett, D., D.C. Bellinger and L.S. Birnbaum. (2016). Project TENDR: Targeting environmental neuro-developmental risks. The TENDR Consensus Statement. *Environmental Health Perspectives, 124*(7), A118–122.

Bennett, H. (Executive producer). (1974–1978). *Six million dollar man* [Television series]. United States: NBC Universal.

Berge, R., and B. Cohen (Producers), and J. Shenk (Director). (2011). *The Island President* [Motion picture]. United States: Samuel Goldwyn Films.

Bergmann, M., S. Mützel, S. Primpke, M.B. Tekman and G. Gerdts. (2018). *Blown to the North? Microplastic in snow fallen out from the atmosphere of Europe and the Arctic.* Presented at MICRO 2018, Lanzarote, Spain. Retrieved from https://epic.awi.de/id/eprint/48513/.

Berners-Lee, T., J. Hendler and O. Lassila, (2001, May). The semantic web. *Scientific American, 284*(5), 34–43.

Bevis, M., C. Harig, S.A. Khan, A. Brown, F.J. Simons, M. Willis ... and D.J. Caccamise. (2019). Accelerating changes in ice mass within Greenland, and the ice sheet's sensitivity to atmospheric forcing. *Proceedings of the National Academy of Sciences, 116*(6), 1934–1939.

Bi, H. (2018). Technique discussion of CRISPR babies-A comment to Jiankui He's research. *Journal of Medical Discovery, 3*(4), 1–2.

Bilalic, M., and P. McLeod. (2014, March). Why good thoughts block better ones. *Scientific American, 310*(3), 74–79.

Billings, L. (2014). The sixth extinction. *Scientific American, 310*(2), 76–76.

Bitcoin Exchange Guide. (2019). *Bitcoin scams and cryptocurrency hacks list.* Retrieved from https://bitcoinexchangeguide.com/bitcoin/scams-hacks/.

BitInfoCharts. (2019). *Bitcoin rich list: Bitcoin distribution.* Retrieved on May 21, 2019 from https://bitinfocharts.com/top-100-richest-bitcoin-addresses.html.

Bittman, M. (2019). *Iowa crops look like food—but no one's eating.* Retrieved from https://www.grain.org/en/article/6291-iowa-crops-look-like-food-but-no-one-s-eating.

Blair, B.G. (1993). *The logic of accidental nuclear war.* Washington, D.C.: Brookings Institution Press.

Blair, B.G., H.A. Feiveson and F.N. von Hippel. (1997, November). Taking nuclear weapons off hair-trigger alert. *Scientific American, 277*(5), 74–81.

Blake, M., and M. Mar. (2014, March). The scary new evidence on BPA-free plastics. *Mother Jones*. Retrieved from https://www.motherjones.com/environment/2014/03/tritan-certichem-eastman-bpa-free-plastic-safe/.

Blanck, H.M., M. Marcus, P.E. Tolbert, C. Rubin, A.K. Henderson, V.S. Hertzberg ... and L. Cameron. (2000). Age at menarche and tanner stage in girls exposed in utero and postnatally to polybrominated biphenyl. *Epidemiology, 11*(6), 641–647.

Blankespoor, B., S. Dasgupta and B. Laplante. (2014). Sea-level rise and coastal wetlands. *Ambio, 43*(8), 996–1005.

Blaser, M. (2014). *Missing microbes: how the overuse of antibiotics is fueling our modern plagues*. New York: Henry Holt.

Blaustein, A.R., and D.B. Wake. (1995, April). The puzzle of declining amphibian populations. *Scientific American, 272*(4), 52–57.

Blaustein, A.R., and P.T. Johnson. (2003). The complexity of deformed amphibians. *Frontiers in Ecology and the Environment, 1*(2), 87–94.

Blaustein, A.R., and P.T.J. Johnson. (2003, February). Explaining frog deformities. *Scientific American, 288*(2), 60–65.

Böhme, R., N. Christin, B. Edelman and T. Moore. (2015). Bitcoin: Economics, technology, and governance. *Journal of Economic Perspectives, 29*(2), 213–238.

Borenstein, S. (2014, September 30). Climate change tied to extreme weather in 2013. *Atlanta Journal-Constitution*. p. A5.

Bottome, E.M. (1971). *Balance of Terror*. Elk Grove Village, IL: Beacon Books.

Bourzac, K. (2014, July). Graphene's dark side. *Scientific American, 311*(1), 19.

Bové, H., E. Bongaerts, E. Slenders, E.M. Bijnens, N.D. Saenen, W. Gyselaers ... and T.S. Nawrot. (2019). Ambient black carbon particles reach the fetal side of human placenta. *Nature Communications, 10*(1), 1–7.

Bowman v. Monsanto Corporation, 569 U.S. 278 (2013).

Boylan, K.D., and D.R. MacLean. (1997). Linking species loss with wetlands loss. *National Wetlands Newsletter, 19*(6), 13–17.

Bradbury, R. (1950). *The Martian chronicles*. Garden City, NY: Doubleday.

Brenig, C., R. Accorsi and G. Müller. (2015, May). Economic analysis of cryptocurrency backed money laundering. *ECIS 2015 Completed Research Papers*, paper 20.

British Broadcasting Corporation (BBC). (2005, January 25). Indonesia quake toll jumps again. *BBC News*. Retrieved on December 24, 2012 from http://news.bbc.co.uk/2/hi/asia-pacific/4204385.stm.

Broad, W.J. (2018, September 2). Microwave weapon might have disabled U.S. diplomats. *Atlanta Journal-Constitution*, p. A15.

Brombacher, A.C. (2006). The impact of new technology on the reliability of future systems: Some food for thought. *Quality and Reliability Engineering International, 22*(4), 369–369.

Brown, E. (2012, July 5). Forbes: Finding the Higgs Boson cost $13.25 billion. *International Business Times*. Retrieved from https://www.ibtimes.com/forbes-finding-higgs-boson-cost-1325-billion-721503.

Brown, J. (2014, June 13). Companies tracking our online footsteps should be more transparent, says FTC (Interview with FTC Chair, Edith Ramirez). *PBS NewsHour*.

Brown, J. (2015, December 14). Italian olive trees are withering from this deadly bacteria. *PBS NewsHour*.

Brumfiel, G. (2013, January 16). Fukushima: Fallout of fear. *Nature, 493*(7432).

Brumfiel, G. (2013, September). Replaceable you. *Smithsonian Magazine, 44*(5), 76–88.

Brundage, M., et al. (2018). The malicious use of artificial intelligence: Forecasting, prevention, and mitigation. *Future of Humanity Institute*. Retrieved from https://maliciousaireport.com/.

Brynjolfsson, E., and A. McAfee. (2014). *The second machine age: Work, progress, and prosperity in a time of brilliant technologies*. New York: WW Norton.

Brzezinski, Z. (1982). *Between two ages: America's role in the technetronic era*. Westport, CT: Praeger.

Buesseler, K. (2012). Fishing for answers off Fukushima. *Science, 338*(6106), 480–482.

Buesseler, K. (2014). Fukushima and ocean radioactivity. *Oceanography, 27*(1), 92–105.

Buesseler, K., M. Dai, M. Aoyama, C. Benitez-Nelson, S. Charmasson, K. Higley ... and J.N. Smith. (2017). Fukushima Daiichi–derived radionuclides in the ocean: Transport, fate, and impacts. *Annual Review of Marine Science, 9*, 173–203.

Bull, A., and J. Finkle. (2013, February 6). Fed says internal site breached by hackers; No critical functions affected. *Reuters.* Retrieved from https://www.reuters.com/article/net-us-usa-fed-hackers/fed-says-internal-site-breached-by-hackers-no-critical-functions-affected-idUSBRE91501920130206.

Bulletin of the Atomic Scientists. (2014). World Nuclear Weapons Stockpiles. https://thebulletin.org/2014/09/worldwide-deployments-of-nuclear-weapons-2014/.

Buolamwini, J., and T. Gebru. (2018, January). Gender shades: Intersectional accuracy disparities in commercial gender classification. *Proceedings of Machine Learning Research, 81*, 77–91.

Burrell, T. (2020). Researchers react to human genes in monkeys. *Discover, 41*(1), 26–27.

Byers, F.R. (2003). Care and handling of CDs and DVDs—A guide for librarians and archivists. *NIST Special Publication, 500*(252), 1–40.

Byrd, D. (2017, May 6). *How long to travel to Alpha Centauri?* Retrieved from https://earthsky.org/space/alpha-centauri-travel-time.

Calderone, J. (2015, April 22). These five ideas will either fix global warming or break the planet. *Reuters.* Retrieved from http://blogs.reuters.com/great-debate/2015/04/22/five-quick-fixes-to-global-warming-and-why-we-probably-shouldnt-try-them/.

Calderone, N.W. (2012). Insect pollinated crops, insect pollinators and US agriculture: Trend analysis of aggregate data for the period 1992–2009. *PLOS One, 7*(5), e37235.

Calderon-Garciduefias, L., and R. Villarreal-Rios. (2017). Living close to heavy traffic roads, air pollution, and dementia. *The Lancet, 389*(10070), 675-677.

Capek, K. (1920). *R.U.R.: Rossum's universal robots.* Prague: Aventinum.

Caravanos, J., K. Chatham-Stephens, B. Ericson, P.J. Landrigan and R. Fuller. (2013). The burden of disease from pediatric lead exposure at hazardous waste sites in 7 Asian countries. *Environmental Research, 120*, 119–125.

Carlsen, E., A. Giwercman, N. Keiding N.E. Skakkebæk. (1992). Evidence for decreasing quality of semen during past 50 years. *BMJ, 305*(6854), 609–613.

Carson, R. (1962). *Silent spring.* New York: Houghton Mifflin.

Castelvecchi, D. (2019, April 4). AI pioneer: "The dangers of abuse are very real." *Nature.* Retrieved from https://www.nature.com/articles/d41586-019-00505-2.

Center for Green Schools. (2013). *State of our schools report.* Retrieved from https://centerforgreenschools.org/sites/default/files/resource-files/2013%20State%20of%20Our%20Schools%20Report%20FINAL.pdf.

Centers for Disease Control and Prevention. (2013). *Antibiotic resistance threats in the United States, 2013.* Retrieved from www.cdc.gov/drugresistance/threat-report-2013/.

Chafkin, M. (2016, September 1). Shields up! New device promises to ward off unwanted drones. *Bloomberg Businessweek.* Retrieved from https://www.bloomberg.com/news/articles/2016-09-01/shields-up-new-device-promises-to-ward-off-unwanted-drones.

Chang, A. (2015, April 24). Man-made earthquakes jolting U.S., report finds. *Atlanta Journal-Constitution*, p. A8.

Chang, K., M. Isaac., and M. Richtel (2016, September 1). SpaceX rocket explodes at launchpad in Cape Canaveral. *New York Times.* Retrieved from https://www.nytimes.com/2016/09/02/science/spacex-rocket-explosion.html.

Chatterjee, P. (2013, March 12). Big data: The greater good or the invasion of privacy? *The Guardian.*

Chechetka, S.A., Y. Yu, M. Tange and E. Miyako. (2017). Materially engineered artificial pollinators. *Chem, 2*(2), 224–239.

Chen, H., J.C. Kwong, R. Copes, K. Tu, P.J. Villeneuve, A. van Donkelaar ... and A.S. Wilton. (2017). Living near major roads and the incidence of dementia, Parkinson's disease, and multiple sclerosis: A population-based cohort study. *The Lancet, 389*(10070), 718–726.

Cheng, M. (2016, February 1). Britain approves contentious gene-editing experiments. *Durango Herald.* Retrieved on February 2, 2016 from https://durangoherald.com/articles/101024.

Christensen, J., T.K. Grønborg, M.J. Sørensen, D. Schendel, E.T. Parner, L.H. Pedersen and M. Vestergaard. (2013). Prenatal valproate exposure and risk of autism spectrum disorders and childhood autism. *Jama*, *309*(16), 1696–1703.

Clark, K. (2013, May). College is free! *Money*, *42*(4), 82–87.

Cloherty, J., and P. Thomas. (2014, November 6). Trojan horse bug lurking in vital US computers since 2011. *ABC News*. Retrieved from https://abcnews.go.com/US/trojan-horse-bug-lurking-vital-us-computers-2011/story?id=26737476.

Cloud, D.S. (2013, November 9). "Iron Man" brought closer to life. *Chicago Tribune*.

Colfield, C. (2015, July 21). *Stephen Hawking: Intelligent aliens could destroy humanity but let's search anyway*. Retrieved from https://www.space.com/29999-stephen-hawking-intelligent-alien-life-danger.html.

Colón, I., D. Caro, C.J. Bourdony and O. Rosario. (2000). Identification of phthalate esters in the serum of young Puerto Rican girls with premature breast development. *Environmental Health Perspectives*, *108*(9), 895.

Comarazamy, D. E., J.E. González, J.C. Luvall, D.L. Rickman and R.D. Bornstein. (2013). Climate impacts of land-cover and land-use changes in tropical islands under conditions of global climate change. *Journal of Climate*, *26*(5), 1535–1550.

Cook, C. (2015). Personal communication.

Cook, J.R. (2012, June). The psychological origins of homophobia and the case of Dharun Ravi. *India America Today*.

Corbett, I. (2012, April). *What data compression does to your music*. Retrieved from http://www.soundonsound.com/techniques/what-data-compression-does-your-music.

Costill, A. (2013, August 7). Top 10 places that have banned google glass. *Search Engine Journal*. Retrieved on September 5, 2014 from https://www.searchenginejournal.com/top-10-places-that-have-banned-google-glass/66585/.

Council on Library Information & Resources. (n.d.). *How long can you store CDs and DVDs and use them again?* Retrieved on August 29, 2014 from http://www.clir.org/pubs/reports/pub121/sec4.html.

Cox, K.D., G.A. Covernton, H.L. Davies, J.F. Dower, F. Juanes and S.E. Dudas. (2019). Human consumption of microplastics. *Environmental Science & Technology*, *53*(12), 7068–7074.

Crane, J. (2010, April 10). UN arms trade treaty receives widespread praise in the CR. *The Prague Post*. Retrieved from http://www.praguepost.com/news/15956-un-arms-trade-treaty-receives-widespread-praisein-cr.html.

Cribb, J. (2014). *Poisoned planet: How constant exposure to man-made chemicals is putting your life at risk*. Sydney, Australia: Allen & Unwin.

Crinnion, W.J. (2000). Environmental medicine, part one: the human burden of environmental toxins and their common health effects. *Alternative Medicine Review*, *5*(1), 52–63.

Croen, L.A., J.K. Grether, C.K. Yoshida, R. Odouli and V. Hendrick. (2011). Antidepressant use during pregnancy and childhood autism spectrum disorders. *Archives of General Psychiatry*, *68* (11), 1 1041112.

Crossman, M. (2016). Iron will. *ESPN*. Retrieved from http://www.espn.com/espn/feature/story/_/page/espnw-kozel161120/quadriplegic-stacey-kozel-conquers-appalachian-trail-exoskeleton.

Cuban, L. (n.d.) Larry Cuban on school reform and classroom practice. https://larrycuban.wordpress.com/.

Cuban, L., H. Kirkpatrick and C. Peck. (2001). High access and low use of technologies in high school classrooms: Explaining an apparent paradox. *American Educational Research Journal*, *38*(4), 813–834.

Cutler, K. (2013, April 3). Another bitcoin wallet service, Instawallet, suffers attack, shuts down until further notice. *TechCrunch*. Retrieved from https://techcrunch.com/2013/04/03/bitcoin-instawallet/.

Dahl, T.E. (1990). *Wetlands losses in the United States, 1780's to 1980's. Report to the Congress (No. PB-91-169284/XAB)*. St. Petersburg, FL: National Wetlands Inventory.

Daly, M. (2013, March 11). Nuclear plants called safer: Watchdog group questions that assertion. *Atlanta Journal-Constitution*, p. B1.

Daly, M. (2016, May 25). House approves bipartisan bill to regulate toxic chemicals. *Atlanta Journal-Constitution*, p. A14.

Darwin, C. (1968/1859). *On the origin of species by means of natural selection*. London: Murray.

Dawson, G. (2013). Dramatic increase in autism prevalence parallels explosion of research into its biology and causes. *JAMA Psychiatry*, *70*(1), 9–10. doi: 10.1001/jama psychiatry.2013.488.

Deeley, M. (Producer), and R. Scott (Director). (1982). Blade Runner [Motion picture]. United States: Warner Bros.

Deere, S. (2018, December 5). Audit: Police violate policy. *Atlanta Journal-Constitution*, p. A1 & A8.

Deeter, J. (Producer), and C. Paine (Director). (2006). *Who killed the electric car?* [Documentary film]. United States: Sony Pictures Classics.

Defense-Aerospace.com. (n.d.) *Maintenance hours for the F-35*. Retrieved from http://www. defense-aerospace.com/.

Den Hond, E., G. Schoeters. (2006). Endocrine disrupters and human puberty. *International Journal of Andrology*, *29*(1), 264–271.

de Laval, B., and M.H. Sieweke. (2017). Trained macrophages support hygiene hypothesis. *Nature Immunology*, *18*(12), 1279.

de Sola Pool, I., and M. Kochen. (1978–1979). Contacts and influence. *Social Networks*, *1*(1), 5–51. Retrieved from https://deepblue.lib.umich.edu/bitstream/handle/2027.42/ 23764/0000737.pdf.

DeSoto, M.C., and R.T. Hitlan. (2007). Blood levels of mercury are related to diagnosis of autism: A reanalysis of an important data set. *Journal of Child Neurology*, *22*(11), 1308–1311.

Diamanti-Kandarakis, E., J.P Bourguignon, L.C. Giudice, R. Hauser, G.S. Prins, A.M. Soto ... and A.C. Gore. (2009, June). Endocrine-disrupting chemicals: An Endocrine Society scientific statement. *Endocrine*, *30*(4), 293–342.

Dick, P.K. (1968). *Do androids dream of electric sheep?* New York: Ballantine Books/Random House.

Dodds, P., R. Muhamad and D.J. Watts. (2003, September). An experimental study of search in global social networks. *Science*, *301*(5634), 827–829.

Dorans, K.S., E.H. Wilker, W. Li, M.B. Rice, P.L. Ljungman, J. Schwartz, and J.M. Massaro. (2016). Residential proximity to major roads, exposure to fine particulate matter, and coronary artery calcium. *Arteriosclerosis, Thrombosis, and Vascular Biology*, *36*, 1679-1685. doi: https://doi.org/10.1161/ATVBAHA.116.307141.

Doyle, A. (2013, August 30). Special report: Experimental climate fixes stir hopes, fears, lawyers. *Reuters*. Retrieved from http://www.retures.com/assets/print?aidUSBRE97TOBZ 20130830.

Doyle, A. (2016, January 19). Killer whales risk extinction off Europe from banned chemicals. *Gwinnett Daily Post*, p. 5A.

Dratva, J., F.G. Real, C. Schindler, U. Ackermann-Liebrich, M.W. Gerbase, N.M. Probst-Hensch and E. Zemp. (2009). Is age at menopause increasing across Europe? Results on age at menopause and determinants from two population-based studies. *Menopause*, *16*(2), 385–394.

Dris, R., H. Imhof, W. Sanchez, J. Gasperi, F. Galgani, B. Tassin, and C. Laforsch. (2015). Beyond the ocean: contamination of freshwater ecosystems with (micro-) plastic particles. *Environmental Chemistry*, *12*(5), 539–550.

Dror, I.E. (2008). Technology enhanced learning: The good, the bad, and the ugly. *Pragmatics & Cognition*, *16*(2), 215–223.

Drysdale, J. (2019/2017). *Our radioactive ocean*. Retrieved from http://www.ourradioactive-ocean.org/.

Du Toit, G., G. Roberts, P.H. Sayre, H.T. Bahnson, S. Radulovic, A.F. Santos ... and V. Turcanu. (2015). Randomized trial of peanut consumption in infants at risk for peanut allergy. *New England Journal of Medicine*, *372*(9), 803–813.

Dubash, M. (2010, April 13). Moore's Law is dead, says Gordon Moore. *Techworld*. Retrieved from https://www.techworld.com/news/tech-innovation/moores-law-is-dead-says-gordon-moore-3576581/.

Dubey, R.B., M. Hanmandlu, and S.K. Gupta. (2010). Risk of brain tumors from wireless phone use. *Journal of Computer Assisted Tomography, 34*(6), 799–807.

Dublon, G., and J.A. Paradiso. (2014). Extra sensory perception. *Scientific American, 311*(1), 36–41.

Dublon, G., J. Paradiso, et. al. (2014). Being there: Fun with a remote sensor data browser (MIT DoppelLab Demonstration). *Scientific American.* Retrieved from https://www.scientificamerican.com/article/sensors-being-there-fun-with-a-remote-sensor-data-browser/.

Dunbar, R.I. (1992). Neocortex size as a constraint on group size in primates. *Journal of Human Evolution, 22*(6), 469–493. Doi:10.1016/0047-2484(92)90081-J.

Dunbar, R.I. (1993). Coevolution of neocortical size, group size and language in humans. *Behavioral and Brain Sciences, 16*(04), 681–694.

Dunbar, R.I. (1995). Neocortex size and group size in primates: A test of the hypothesis. *Journal of Human Evolution, 28*(3), 287–296.

Dunmore, C. (2013, April 29). EU to ban pesticides blamed for harming bees. *Reuters.* Retrieved from https://www.reuters.com/article/us-eu-pesticides/eu-to-ban-pesticides-blamed-for-harming-bees-idUSBRE93S0CX20130429.

Dyer, O. (2018). Microwave weapon caused syndrome in diplomats in Cuba, US medical team believes. *BMJ, 362.* doi: https://doi.org/10.1136/bmj.k3848.

Edenhofer, O., et al., eds. (2014). *IPCC: Climate change 2014: Mitigation of climate change, contribution of Working Group 1 to the fifth assessment report of the Intergovernmental Panel on Climate Change.* Cambridge, England: Cambridge University Press.

Edwards, L. (2010, June 23). Humans will be extinct in 100 years says eminent scientist. *Science X Network.* Retrieved from https://phys.org/news/2010-06-humans-extinct-years-eminent-scientist.html.

Eggers, D. (2013). *The circle.* New York: Knopf.

Ehlinger, S. (2014, November 30). Some $800 million paid to SRS workers for Cold War radiation exposure. https://www.thestate.com/news/politics-government/politics-columns-blogs/the-buzz/article13920077.html.

Eisen, S. (2015, December 25). MIT invention can see through walls: Technology raises questions of privacy rights and intrusion. *Atlanta Journal-Constitution,* pp. A17–18.

Elias, P. (2015, May 20). Six Chinese nationals charged with espionage. *Atlanta Journal-Constitution,* p. A2.

Elias, S.A. (2006). *Encyclopedia of Quaternary Science.* Amsterdam: Elsevier.

Elsaesser, A., and C.V. Howard. (2012). Toxicology of nanoparticles. *Advanced Drug Delivery Reviews, 64*(2), 129–137. doi: 10.1016/j.addr.2011.09.001.

Emmerich, R. (Director), R. Emmerich and M. Gordon (Producers). (2004). *Day After Tomorrow* [Motion picture]. United States: 20th Century Fox/Lion's Gate.

Engelsdorp, D.V., J.S. Pettis, and W. Ritter. (2014). Colony collapse disorder. In *Bee health and veterinarians,* 157–159. Paris: OIE (World Organization for Animal Health).

Environmental Protection Agency. (2016). *EPA's Study of Hydraulic Fracturing for Oil and Gas and Its Potential Impact on Drinking Water Resources.* Retrieved from https://www.epa.gov/hfstudy.

Environmental Working Group. (2007). 455 industrial pollutants in the blood and tissues of Americans. Cited in Savan,(2007).

Epps, L.C., and P.D. Walker. (2006). Fluoroquinolone consumption and emerging resistance. *U.S. Pharmacist, 10,* 47–54.

Estep, P.W., M. Kaeberlein, P. Kapahi, B.K. Kennedy, G.J. Lithgow, G.M. Martin, ... and H.A. Tissenbaum. (2006). Life extension pseudoscience and the SENS plan. *MIT Technology Review, 109*(3), 80–84.

EurActiv with Reuters. (2014, July 16). *"Smart" technology makes energy utilities more vulnerable to hackers.* Retrieved on September 9, 2014 from https://www.euractiv.com/section/energy/news/smart-technology-makes-energy-utilities-more-vulnerable-to-hackers/.

European Space Agency. (2005, April 29). *Earth from space: North Sentinel Island.* Retrieved from https://www.esa.int/Applications/Observing_the_Earth/Earth_from_Space_ North_Sentinel_Island.

Evans, G.W., and E. Kantrowitz. (2002). Socioeconomic status and health: the potential role of environmental risk exposure. *Annual Review of Public Health, 23*(1), 303–331.

Everstine, B. (2015, April 15). Problems plaguing F-35's next-gen maintenance systems. *Airforce Times.* Retrieved from https://www.airforcetimes.com/news/your-air-force/ 2015/04/15/problems-plaguing-f-35-s-next-gen-maintenance-system/.

Face the Nation. (2017, February 6). Islamic state uses off-the shelf drones. *CBS News.*

Face the Nation. (2017, February 26). ISIS uses drones to drop bombs in battle of Mosul. *CBS News.*

Falconer, B. (2014, May 5). Can anyone stop the man who will try just about anything to put an end to climate change? *Pacific Standard.* Retrieved on February 17, 2016 from http://www.psmag.com/booksand-culture/battlefield-ealth-can-anyone-stop-man-will-try-just-anything-fix-climate-78957.

Famiglietti, J.S., A. Cazenave, A. Eicker, J.T. Reager, M. Rodell, and I. Velicogna. (2015). Satellites provide the big picture. *Science, 349*(6249), 684–685.

Famous Women Inventors. (n.d.) *Hedy Lamarr: Invention of spread spectrum technology.* Retrieved on November 19, 2015 from http://www.women-inventors.com/Hedy-Lammar. asp.

Farley, B.W.A.C., and W. Clark. (1954). Simulation of self-organizing systems by digital computer. *Transactions of the IRE Professional Group on Information Theory, 4*(4), 76–84.

Farrar, W. (2014). Operation Candid Camera: Rialto police department's body-worn camera experiment. *The Police Chief, 81,* 20–25.

Farrar, W., and B. Ariel. (2013, March). Self-awareness to being watched and socially-desirable behavior: A field experiment on the effect of body-worn cameras and police use-of-force. *National Police Foundation.* Retrieved from https://www.policefoundation.org/ publication/self-awareness-to-being-watched-and-socially-desirable-behavior-a-field-experiment-on-the-effect-of-body-worn-cameras-on-police-use-of-force/.

Federal Aviation Administration (FAA). (2014). *Laser pointers up 1200%.* Retrieved from https://www.faa.gov/news/updates/?newsId.

Federal Aviation Administration (FAA). (2019). *FAA updates on Boeing 737 MAX.* https:// www.faa.gov/news/updates/?newsId=93206.

Federal Bureau of Investigation (FBI). (2014). Laser pointers incidents. *Uniform Crime Reports.* U.S. Department of Justice.

Fela, J. (2018, Fall). Campaigning for a plastic-free future. Greenpeace.

Fernandez-Cornejo, J., S. Wechsler, M. Livingston, and L. Mitchell. (2014). Genetically engineered crops in the United States. *USDA-ERS Economic Research Report* (162). Retrieved from https://www.ers.usda.gov/webdocs/publications/45179/43668_err162.pdf?v=0.

Field, C.B. et al, (eds.). (2012). *IPCC: Managing the risks of extreme events and disasters to advance climate change adaptation.* Cambridge, England: Cambridge University Press.

Finkle, J. (2014, July 31). Hackers can tap USB devices in new attacks, researcher warns. *Reuters.* https://www.reuters.com/article/us-cybersecurity-usb-attack/hackers-can-tap-usb-devices-in-new-attacks-researcher-warns-idUSKBN0G00K420140731.

First, M.B. (2005). Desire for amputation of a limb: Paraphilia, psychosis, or a new type of identity disorder. *Psychological Medicine, 35*(6), 919–928. doi:10.1017/S0033291704003320.

Fischetti, M. (2010). The great chemical unknown. *Scientific American, 303*(5), 92–92.

Fisher, N.S., K. Beaugelin-Seiller, T.G. Hinton, Z. Baumann, D.J. Madigan and J. Garnier-Laplace. (2013). Evaluation of radiation doses and associated risk from the Fukushima nuclear accident to marine biota and human consumers of seafood. *Proceedings of the National Academy of Sciences, 110*(26), 10670–10675.

Flaherty, A. (2014, December 8). Kids still being tracked by phone apps. *The Press Democrat.* Retrieved from https://www.pressdemocrat.com/home/3222354-181/ kids-still-being-tracked-by.

Flatow, I. (2011, April 22). Science diction: The origin of the word "robot." *National Public Radio.* Retrieved from https://www.npr.org/2011/04/22/135634400/science-diction-the-origin-of-the-word-robot.

Florida Microplastics Awareness Project. (2017). *Microplastics*. University of Florida. Retrieved from https://sfyl.ifas.ufl.edu/flagler/marine-and-coastal/microplastics/.

Foley, R.J., and M. Hoyer. (2016, April 9). U.S. water systems repeatedly exceed federal standard for lead. *Associated Press*. Retrieved from https://apnews.com/5aff8cb852c94585a85c9dc5fa32e9d8/us-water-systems-repeatedly-exceed-federal-standard-lead.

Food and Agriculture Organization of the UN. (2010). *Global Forest Resources Assessment Main Report*. Rome: Food and Agriculture Organization, 378.

Ford, M. (2015). *Rise of the robots: Technology and the threat of a jobless future*. New York: Basic Books.

Foreign Cyber Threats to the United States: Hearings before the Committee on Armed Services, Senate, 115th Cong. (2017).

Fountain, H. (2012, October 19). A rogue climate experiment has ocean experts outraged. *New York Times*, p. Al.

Fox, M. (2014, June 26). ReWalk robotic exoskeleton suit gets FDA approval. *NBC News*. Retrieved from https://www.nbcnews.com/health/health-news/rewalk-robotic-exoskeleton-suit-gets-fda-approval-n142116.

Fredrickson, B. (2013). *Love 2.0: How our supreme emotion affects everything we feel, think, do and become*. New York: Gildan Media.

Free Library (2014). San Onofre power plant home to high-energy kittens. The Free Library. Retrieved from http://www.thefreelibrary.com. Originally published by LA Times at Perry, T. (1996, February 8). Cats cradled in nuclear plant are reportedly healthy. *Los Angeles Times*. https://www.latimes.com/archives/la-xpm-1996-02-08-mn-33724-story.html.

Friedland, A.E., Y.B. Tzur, K.M. Esvelt, M.P. Colaiàcovo, G.M. Church and J.A. Calarco. (2013). Heritable genome editing in C. elegans via a CRISPR-Cas9 system. *Nature Methods*, 10(8), 741–743.

Friends of the Earth. (2015). *Bee action*. Retrieved from http://www.foe.org/projects/food-andtechnology/beeaction.

Friends of the Earth. (2015, summer). Paying the high costs of climate chaos: The story of climate finance. *Friends of the Earth News Magazine*, pp. 16–17.

Friends of the Earth International. (2014). *Annual report*. Retrieved from https://www.foei.org/wp-content/uploads/2015/06/FoEI-Annual-Report-2014.pdf.

Fung, B. (2016, December 30). Amazon envisions flying warehouses. *Atlanta Journal-Constitution*, p. A11.

Furnham, A. (2003). Belief in a just world: Research progress over the past decade. *Personality and Individual Differences*, 34, 795–817.

Gallup, G.G. (1970). Chimpanzees: Self-recognition. *Science*, 167(3914), 86–87.

Gàmez-Guadix, M., I. Orue, P.K. Smith and E. Calvete. (2013). Longitudinal and reciprocal relations of cyberbullying with depression, substance use, and problematic internet use among adolescents. *Journal of Adolescent Health*, 53(4), 446–452.

Garamone, J. (2014, January 15). 34 ICBM launch officers implicated in cheating probe. *United States Air Force*. Retrieved from https://www.af.mil/News/Article-Display/Article/468806/34-icbm-launch-officers-implicated-in-cheating-probe/.

Garamszegi, L.Z. (2011). Climate change increases the risk of malaria in birds. *Global Change Biology*, 17(5), 1751–1759.

Garfinkle, D., J. Renfroe, S. Rankin, D. Contis and J. Boyle (Executive producers), and K. Hunter and A. Shaffir (Directors). (2013–present). *Naked & Afraid* [Television series]. United States: Discovery Channel.

Gartner. (2016, May 19). Gartner says worldwide smart phone sales grew 3.9 percent in first quarter of 2016. https://www.gartner.com/en/newsroom/press-releases/2016-05-19-gartner-says-worldwide-smartphone-sales-grew-4-percent-in-first-quarter-of-2016.

Gates, W. (2007, January). A robot in every home. *Scientific American*, 296(1), pp. 58–65.

Gelineau, K. (2015, December 10). Bitcoin "inventor's" residence raided. *Atlanta Journal-Constitution*, p. A17.

Gellman, B., and A. Soltani. (2013, October 30). NSA infiltrates links to Yahoo, Google data centers worldwide, Snowden documents say. *The Washington Post*.

Gendron, B. (1977). *Technology and the human condition*. New York: St. Martin's Press.

Genoways, T. (2013, July/August). Gagged by big ag. *Mother Jones.*

GeoEngineering Watch. (2015). Retrieved from www.geoengineeringwatch.org.

GermanWatch. (2004). *Sea-level rise in The Netherlands and Bangladesh: One phenomenon, many consequences.* Retrieved on September 9, 2014 from https://germanwatch.org/sites/germanwatch.org/files/publication/3642.pdf.

Gibbs, J.P. (2000). Wetland loss and biodiversity conservation. *Conservation Biology, 14*(1), 314–317.

Gibson, J. F., E.D. Stein, D.J. Baird, C.M. Finlayson, X. Zhang and M. Hajibabaei. (2015). Wetland ecogenomics—the next generation of wetland biodiversity and functional assessment. *Wetland Science and Practice, 32,* 27–32.

Gill, V. (2010, May 20). "Artificial life," breakthrough announced by scientists. *BBC News.* Retrieved on February 10, 2015 from www.bbc.com/news/10132762.

Gillum, J., and E. Sullivan. (2014, June 13). U.S.: Keep data sweeps secret. *Atlanta Journal-Constitution,* p. A5.

Glasser, M.F., S.M. Smith, D.S. Marcus, J.L. Andersson, E.J. Auerbach, T.E. Behrens and E.C. Robinson. (2016). The Human Connectome Project's neuroimaging approach. *Nature Neuroscience, 19*(9), 1175–1187.

Global Agriculture. (2015). *Special Report: Water.* Retrieved from http://www.globalagriculture.org/report-topics/water.html.

Global Alliance on Health and Pollution. (2013). *The poisoned poor: Toxic chemicals exposures in low and middle-income countries.* Retrieved from https://gahp.net/wp-content/uploads/2017/02/GAHPPoisonedPoor_Report-Sept-2013.pdf.

Global Ocean Commission. (2015). *Summary report: The global ocean.* Retrieved from http://ec.europa.eu/dgs/maritimeaffairs_fisheries/consultations/oceangovernance/contributions/doc/global-ocean-commission-report_en.pdf.

The Globalist. (2013, December 24). 20 facts about Mikhail Kalashnikov's AK-47. *The Globalist.* Retrieved from https://www.theglobalist.com/20-facts-mikhail-kalashnikov-ak-47/.

The Grand Illusion. (2015). *Monsanto vs. farmer.* Retrieved from https://thegrandillusion.wordpress.com/monsantovs.-farmer.

Goldberg, L., R. Hashimoto, H. Schneider, and B. McNall (Producers), and Badham, J. (Director). (1983). *War games* [Motion Picture]. United States: MGM.

Goldman, L, and N. Tran. (2002, August). *Toxins and poverty: The impact of toxic substances on the poor in developing countries.* World Bank. Retrieved on January 5, 2015 from http://documents.worldbank.org/curated/en/689811468315541722/Toxics-and-poverty-the-impact-of-toxic-substances-on-the-poor-in-developing-countries.

Goldman, R. (2017, February 20). Dubai plans taxi that skips driver—and roads. *Atlanta Journal-Constitution,* p. A10.

Goldstein, J. (2013, October 2). Arrest in U.S. shuts down a black market for narcotics. *New York Times,* p. A3.

Goldstone, J. (Producer), and T. Jones (Director). (1983). *Monty Python*: The meaning of life [Motion picture]. United Kingdom: Universal Pictures.

Gonzales, J. (2014, June 25). Police won't divulge tracking tech. *Wilmington Star News,* p. Bl & B5.

Gonzalez, B. (2017). Drones and privacy in the golden state. *Santa Clara High Technology Law Journal, 33*(2), 288.

Goodall, J. (2010). *In the shadow of man.* New York: Mariner Books.

Goodman, L.M. (2014, March 6). The face behind bitcoin. *Newsweek.*

Goold, B. (2018). Thinking ethically about public area surveillance. In *The Routledge Handbook of Ethics and Public Policy.* Abingdon-on-Thames, England, UK: Routledge.

Gordon, R.J. (2016). *The rise and fall of American growth: The US standard of living since the civil war.* Princeton, NJ: Princeton University Press.

Gorgone, J., G.B. Davis, J.S. Valacich, H. Topi, D.L. Feinstein and H.E. Longenecker. (2003). IS 2002 model curriculum and guidelines for undergraduate degree programs in information systems. *Communications of the Association for Information Systems, 11*(1), I.

Gosk, S. (2014, October 9). How safe is artificial turf? *MSNBC* (Morning Joe).

Gould, S.J. (2002). *The structure of evolutionary theory.* Cambridge, MA: Belknap Press.

Graham, U.M., G. Jacobs, R.A. Yokel, B.H. Davis, A.K. Dozier, M.E. Birch... and L. DeLouise.

(2017). From dose to response: In vivo nanoparticle processing and potential toxicity. In *Modelling the Toxicity of Nanoparticles*, pp. 71–100. New York: Springer Nature.

The Grand Illusion. (2015). *Monsanto vs. farmer.* Retrieved from https://thegrandillusion. wordpress.com/monsantovs.-farmer.

Greenberg, A. (2013, June 2). Researchers say they can hack your iPhone with a malicious charger. *Forbes.* Retrieved on November 10, 2015 from www.forbes.com/sites/andygreenberg; 2013/06/02/researchers-say-they-can-hack-your-iPhone-withmalicious-charger.

Greenpeace. (2017, winter). Greenpeace report exposes risks of spills. *Compass,* p. 11.

Greenwald, G. (2013, June 6). NSA collecting phone records of millions of Verizon customers daily. *The Guardian.* Retrieved from https://www.theguardian.com/world/2013/jun/06/ nsa-phone-records-verizon-court-order.

Greenwald, G. (2014). *No place to hide: Edward Snowden, the NSA, and the US surveillance state.* New York: Macmillan.

Griffin, A. (2014, June 8). Turing test breakthrough as super-computer becomes first to convince us it's human. *The Independent.* Retrieved from https://www.independent.co.uk/ life-style/gadgets-and-tech/computer-becomes-first-to-pass-turing-test-in-artificial-in-telligence-milestone-but-academics-warn-9508370.html.

Grossman, L. (2014, February 17). The infinity machine. *Time.*

Grossman, L., and J. Newton-Small. (2013, November 11). The deep web. *Time, 182*(20), pp. 26–33.

Grubin, D., and B. Holman (Producers), and D. Grubin (Director). (2015). *Language matters with Bob Holman* [Documentary film]. United States: Public Broadcasting Service (PBS).

Guinness World Records. (2019). *Sports Records.* Retrieved from www.guinnessworldrecords. com.

Gymrek, M., A.L. McGuire, D. Golan, E. Halperin and Y. Erlich. (2013). Identifying personal genomes by surname inference. *Science, 339*(6117), 321–324.

Hadhazy, A. (2010, May 4). Zettabytes now needed to describe global data overload. *Live Science.*

Hallmann, C.A., M. Sorg, E. Jongejans, H. Siepel, N. Hofland, H. Schwan ... and D. Goulson. (2017). More than 75 percent decline over 27 years in total flying insect biomass in protected areas. *PLOS One, 12*(10), e0185809.

Hammer, J. (2013, March). Lost tribes of the Amazon. *Smithsonian Magazine, 43*(11), 38–49.

Hanna, J. (2014, May 13). Ice melt in part of Antarctica "appears unstoppable," NASA says. *CNN.*

Hanna, W., and J. Barbera (Directors). (1962–1968). *The Jetsons* [Television series]. United States: Hanna-Barbera Productions.

Hardin, G. (1968). The tragedy of the commons. *Science, 162*(3859), 1243–1248. doi: 10.1126/ science.162.3859.1243.

Harkinson, J. (2014, December). Down in the hole. *Mother Jones,* pp. 48–53.

Harris, M. (2016, January 13). Google reports self-driving car mistakes: 272 failures and 13 near misses. *The Guardian.* Retrieved from www.theguardian.com/technology/2016/ jan/12/google-selfdriving-cars-mistakes-data-reports.

Harrison J.W., and T.A. Svec. (1998, April). The beginning of the end of the antibiotic era? Part II. Proposed solutions to antibiotic abuse. *Quintessence International, 29*(4), 223–229.

Hart, M.H. (1974). Explanation for the lack of extraterrestrials on earth. *Quarterly Journal of the Royal Astronomical Society, 16,* 128–135. Retrieved from http://adsabs.harvard.edu/ full/1975QJRAS..16..128H.

Hartman, B. (2015, December 28). Two Israelis arrested for flying drone over Vatican. *The Jerusalem Post.* Retrieved from https://www.jpost.com/Israel-News/Two-Israelis-arrested-for-flying-drone-over-Vatican-438628.

Hartmann, T. (2004). *The last hours of ancient sunlight.* New York: Potter.

Hassan, A. (Ed.). (2015). *Everyday environmental toxins: Children's exposure risks.* Palm Bay, FL: Apple Academic Press.

Hawking, S. (Creator). (2014, May 21). *Brave new world* [Television series]. Designer human [Series episode]. United States: Discover Channel.

Hawks J., K. Hunley, S.H. Lee, M. Wolpoff. (January 2000). Population bottlenecks and

Pleistocene human evolution. *Molecular Biology and Evolution, 17*(1), 2–22. doi: 10.1093/oxfordjournals.molbev.a026233.

Hay, C.C., E. Morrow, R.E. Kopp and J.X. Mitrovica. (2015). Probabilistic reanalysis of twentieth century sea-level rise. *Nature, 517*(7535), 481–484.

Hayflick, L. (1965). The limited in vitro lifetime of human diploid cell strains. *Experimental Cell Research, 37*(3), 614–636.

Hayflick, L. (2000). The illusion of cell immortality. *British Journal of Cancer, 83*(7), 841.

Haynes, A.B., T.G. Weiser, W.R. Berry, S.R. Lipsitz, A.H.S. Breizat, E.P. Dellinger and A.F. Merry. (2009). A surgical safety checklist to reduce morbidity and mortality in a global population. *New England Journal of Medicine, 360*(5), 491–499.

Heberer, T. (2002). Tracking persistent pharmaceutical residues from municipal sewage to drinking water. *Journal of Hydrology, 266*(3), 175–189.

Heitman, M. (Producer). (2011). *Curiosity* [Television series]. I, caveman [Series episode]. United States: Discovery Channel.

Henaghan, M. (2014). Indigenous people, emerging research, and global health. In *Law and Global Health, Current Legal Issues,* Volume 16, 182–192. Oxford University Press. Oxford, UK.

Hendryx, M., and M.M. Ahern. (2009). Mortality in Appalachian coal mining regions: The value of statistical life lost. *Public Health Reports, 124*(4), 541–550.

Henein, M., and G. Langworthy (Directors). (2009). *Vanishing of the bees* [Documentary film]. United Kingdom: Hipfuel Films and Hive Mentality Productions.

Henry, R. (2014, July 28). Nuclear plants challenge builders. *Atlanta Journal-Constitution,* p. B3.

Henry, R. (2015, May 17). 1970s-era plant nears completion. *Atlanta Journal-Constitution,* p. A10.

Herbert, M.R. (2010). Contributions of the environment and environmentally vulnerable physiology to autism spectrum disorders. *Current Opinion in Neurology, 23*(2), 103–110.

Hern, A. (2018, January 28). Fitness tracking app Strava gives away location of secret US army bases. *The Guardian.* Available at https://www.theguardian.com/world/2018/jan/28/fitness-tracking-app-gives-away-location-of-secret-us-army-bases.

Hertz-Picciotto, I., L.A. Croen, R. Hansen, C.R. Jones, J. van de Water and I.N. Pessah. (2006). The CHARGE study: An epidemiological investigation of genetic and environmental factors contributing to autism. *Environmental Health Perspectives, 1,* 119–1126.

Hertz-Picciotto, I., R.J. Schmidt and P. Krakowiak. (2018). Understanding environmental contributions to autism: Causal concepts and the state of science. *Autism Research, 11*(4), 554–586.

Hodge, M.R., W. Horton, T. Brown, R. Herrick, T. Olsen, M.E. Hileman and G.C. Burgess. (2016). ConnectomeDB—sharing human brain connectivity data. *Neuroimage, 124,* 1102–1107.

Hoffer, E. (1996). *The ordeal of change.* Cutchogue, NY: Buccaneer Books.

Hoffman, J. (2012). *Geology and human health: Potential health and environmental effects of hydrofracking in the Williston Basin, Montana.* Retrieved on November 3, 2015 from https://serc.carleton.edu/NAGTWorkshops/health/case_studies/hydrofracking_w.html.

Horsetalk. (2007, August 22) *The costs of owning a horse.* Available from https://horsetalk.co.nz/features/horsecosts-126.shtml#axzz3SCw9Ipdy.

Howard, B.C. (n.d.). Mighty rivers run dry from overuse. *National Geographic.* Retrieved on November 3, 2015 from http://environment.nationalgeographic.com/environment/photos/rivers-run-dry/.

Huber, R., V. Treyer, A.A. Borbely, J. Schuderer, J.M. Gottselig, H.P. Landolt, E. Werth, T. Berthold, N. Kuster, A. Buck, and P. Achermann, (2002, December) Electromagnetic fields, such as those from mobile phones, alter regional cerebral blood flow and sleep and waking EEG. *Journal of Sleep Research, 11*(4), 289–295.

Hughes, J.J. (2012). The politics of transhumanism and the techno-millennial imagination. *Zygon, 47*(4), 757–776.

Hughes, T.P., J.T. Kerry, M. Alvarez-Noriega, J.G. Álvarez-Romero, K.D. Anderson, A.H. Baird and T.C. Bridge. (2017). Global warming and recurrent mass bleaching of corals. *Nature, 543*(7645), 373–377. doi:10.1038/nature21707.

Hulme, M. (2009). On the origin of "the greenhouse effect": John Tyndall's 1859 interrogation of nature. *Weather, 64*(5), 121–123.

Human Rights Watch (2016, April 11). Killer robots and the concept of meaningful human control. Retrieved from https://www.hrw.org/news/2016/04/11/killer-robots-and-concept-meaningful-human-control.

Hutchison, C.A., R.Y. Chuang, V.N. Noskov, N. Assad-Garcia, T.J. Deerinck, M.H. Ellisman ... and J.F. Pelletier. (2016). Design and synthesis of a minimal bacterial genome. *Science, 351*(6280), aad6253.

Innocence Project. (2019). *DNA exoneree case profiles.* Retrieved on May 16, 2019 from https://www.innocenceproject.org/dna-exonerations-in-the-united-states/.

Intergovernmental Panel on Climate Change (IPCC). (2001). *Climate change 2001: Impacts, adaptation and vulnerability.* Cambridge, England: Cambridge University Press.

Intergovernmental Panel on Climate Change (IPCC). (2018, October). *Special report: Global warming of 1.5 degrees C.* Switzerland: Intergovernmental Panel on Climate Change.

Intergovernmental Panel on Climate Change (IPCC). (2019). *Special report: Climate change and land.* Cambridge, England: Cambridge University Press.

Intergovernmental Science-Policy Platform on Biodiversity and Ecosystem Services (IPBES). (2019). *IPBES global assessment report on biodiversity and ecosystem services.* Retrieved from https://ipbes.net/global-assessment.

International Atomic Energy Agency (IAEA). (2011, November 17). *Computer security at nuclear facilities; IAEA nuclear security series.* Retrieved from https://www.iaea.org/publications/8691/computer-security-at-nuclear-facilities.

International Atomic Energy Agency (IAEA). (2014). *Nations with nuclear weapons.* Retrieved from https://www.iaea.org/.

International Atomic Energy Agency (IAEA). (2014). *Nuclear share of electricity generation 2013.* Retrieved from https://www.iaea.org/.

International Atomic Energy Agency. (2018). *Nuclear reactors in the world.* Retrieved from https://www.iaea.org/publications/13379/nuclear-power-reactors-in-the-world.

International Atomic Energy Agency (IAEA). (2019, March 12). *Report on the discharge record and the seawater monitoring results at Fukushima Daiichi Nuclear Power Station during February.* Retrieved from https://www.iaea.org/sites/default/files/19/03/fukushima_nps_update_-_2019-03-12.pdf.

International Energy Agency (IEA). (2015). *Energy access database: World energy outlook.* Retrieved from http://www.worldenergyoutlook.org/resources/energydevelopment/energyaccessdatabase/.

International Energy Agency (IEA). (2016, November). *World Energy Outlook 2016.* Retrieved from https://www.iea.org/reports/world-energy-outlook-2016.

International Global Markets (IGM) Forum. (2013). *Chicago Booth School of Economics: Bitcoin poll.* Retrieved from http://www.igmchicago.org/igm-economic-experts-panel/pollresults?SurveyID=SV8qRwhHaLc7b5Sp7.

International Telecommunications Union. (2016). *Measuring the information society report.* Retrieved from https://www.itu.int/en/ITU-D/Statistics/Documents/publications/misr2016/MISR2016-w4.pdf.

IPCC. (2014a). Climate change 2014: Synthesis report of the fifth assessment report of the IPCC.

Jeffries, A. (2012, August 27). Suspected multi-million-dollar Bitcoin pyramid scheme shuts down, investors revolt. *The Verge.* Retrieved from https://www.theverge.com/2012/8/27/3271637/bitcoin-savings-trust-pyramid-scheme-shuts-down.

Jepson, P.D., R.E. Deaville, J.L. Barber, A. Aguilar, A. Borrell, S. Murphy... and A.A. Cunningham. (2016, January). PCB pollution continues to impact populations of orcas and other dolphins in European waters. *Scientific Reports, 6.* doi: 10.1038/srep18573.

Johnson, K. (Creator). (1976-1978). *Bionic Woman* [Television Series]. United States: MCA/Universal/Harve Bennett Productions.

Johnson, N., L. Reidy, M. Droll and R.E. LeMon. (2012, July). Program requirements for associate's and bachelor's degrees: A national survey. *Complete College America.* Retrieved from https://www.insidehighered.com/sites/default/server_files/files/Program%20Requirements%20-%20A%20National%20Survey%281%29.pdf.

Johnson, R. (2013, December 31). Impact of last year's rouge ocean fertilization experiment still unclear. *Earth Island Journal*. Retrieved February 17, 2016 from https://www.earthisland.org/journal/index.php/articles/entry/impact_of_last_years_rouge_ocean_fertilization_experiment_still_unclear/.

Johnson, S. (2018). *How we got to now: Six innovations that made the modern world.* New York: Viking Books.

Johnston, K., M. Tanner, N. Laila and D. Kawaiski. (2013). *Social capital: The benefit of the Joint Monitoring Program.* Joint Monitoring Program of WHO and UNICEF on Water Supply and Sanitation.

Jolly, W.M., M.A. Cochrane, P.H. Freeborn, Z.A. Holden, T.J. Brown, G.J. Williamson and D.M. Bowman. (2015). Climate-induced variations in global wildfire danger from 1979 to 2013. *Nature Communications, 6,* 7537.

Jones, C. (2016, June 6). Scientists say we're losing battle to save world's coral reefs. *Atlanta Journal-Constitution,* p. A4.

Jones, K.R., C.J. Klein, B.S. Halpern, O. Venter, H. Grantham, C.D. Kuempel ... and J.E. Watson. (2018). The location and protection status of Earth's diminishing marine wilderness. *Current Biology, 28*(15), 2506–2512.

Jones, O.A., J.N. Lester and N. Voulvoulis. (2005). Pharmaceuticals: A threat to drinking water? *Trends in Biotechnology, 23*(4), 163–167.

Kageyama, Y. (2016, September 12). Japan's ex-leader raises funds for ill U.S. troops. *Atlanta Journal-Constitution,* p. A5.

Karinthy. F. (1929). Chain-Links. In *Everything is different (Fifty-Two Sundays).* Budapest: Athenaeum. (A. Makkai, Trans. and E. Jankö, Ed. Retrieved from https://djjr-courses.wdfiles.com/local--files/soc180%3Akarinthy-chain-links/Karinthy-Chain-Links_1929.pdf).

Kaye, B., and J. Wagstaff. (2016, May 2) Australian tells BBC he created bitcoin, but some skeptical. *Reuters.*

Kearney, M., W.P. Porter, C. Williams, S. Ritchie and A.A. Hoffmann. (2009). Integrating biophysical models and evolutionary theory to predict climatic impacts on species' ranges: The dengue mosquito Aedes aegypti in Australia. *Functional Ecology, 23*(3), 528–538.

Kerlik, B. (2014, June 8). FBI campaign aims to raise awareness about dangers of pointing lasers at aircraft. *Pittsburgh Tribune-Review.* Available from https://archive.triblive.com/local/pittsburgh-allegheny/fbi-campaign-aims-to-raise-awareness-about-dangers-of-pointing-lasers-at-aircraft/#ixzz3EzwPqzUr.

Kern, J.K., D.A. Geier, L.K. Sykes, B.E. Haley and M.R. Geier. (2016). The relationship between mercury and autism: A comprehensive review and discussion. *Journal of Trace Elements in Medicine and Biology, 37,* 8–24.

Khan, A. (2016, October 14). Radiation may leave astronauts with "space brain." *Atlanta Journal-Constitution,* p. A9.

Khan, A. (2017, February 20). As bee populations dwindle, drone pollinators' use studied. *Atlanta Journal-Constitution,* p. A14.

Klein, N. (2014). *This changes everything: Capitalism vs. the climate.* New York: Simon and Schuster.

Kleinlogel, H., T. Dierks, T. Koenig, H. Lehmann, A. Minder and R. Berz. (2008). Effects of weak mobile phone—electromagnetic fields (GSM, UMTS) on event related potentials and cognitive functions. *Bioelectromagnetics, 29*(6), 488–497.

Knickmeyer, E. (2019, July 14). EPA restores wide use of pesticide opposed by beekeepers. *Atlanta Journal-Constitution,* p. A7.

Koeppel, D. (2008). *Banana: The fate of the fruit that changed the world.* New York: Penguin.

Koeppel, D. (2011, July 22). The beginning of the end for bananas? *The Scientist.* Retrieved from https://www.the-scientist.com/news-opinion/the-beginning-of-the-end-for-bananas-42182.

Kok, B.E., K.A. Coffey, M.A. Cohn, L.I. Catalino, T. Vacharkulksemsuk, S.B. Algoe, M. Brantley and B.L. Fredrickson. (2013). How positive emotions build physical health: Perceived positive social connections account for the upward spiral between positive emotions and vagal tone. *Psychological Science, 24*(7), 1123–1132. doi: 10.1177/0956797612470827.

Kolbert, E. (2015). *The sixth extinction: An unnatural history*. New York: Henry Holt.

Kosinski, M., D. Stillwell and T. Graepel. (2013). Private traits and attributes are predictable from digital records of human behavior. *Proceedings of the National Academy of Sciences, 110*(15), 5802–5805.

Kosinski, M., Y. Wang, H. Lakkaraju and J. Leskovec. (2016). Mining big data to extract patterns and predict real-life outcomes. *Psychological Methods, 21*(4), 493.

Kramer, A.E. (2017, January 28). Arrest in Russia linked to election hacking. *Atlanta Journal-Constitution*, p. A8.

Kramer, A.E. (2019, August 12). Russia confirms radioactive materials involved in blast. *Atlanta Journal-Constitution*, p. A8.

Krapp, M., R. Beyer, S.L. Edmundson, P.J. Valdes and A. Manica. (2019, August). A comprehensive history of climate and habitat stability of the past 800 thousand years. *Climate of the Past Discussions*.

Kristensen, H.M., and R.S. Norris. (2017). Worldwide deployments of nuclear weapons, 2017. *Bulletin of the Atomic Scientists, 73*(5), 289–297.

Krolicki, K., & Fujioka, C. (2011). Special report: Japan's "throwaway" nuclear workers. Retrieved from https://www.reuters.com/article/us-japan-nuclear-idUSTRE75N18A 20110624.

Kruger, J., and D. Dunning. (1999). Unskilled and unaware of it: How difficulties in recognizing one's own incompetence lead to inflated self-assessments. *Journal of Personality and Social Psychology, 77*(6), 1121. doi:10.1037/0022–3514.77.6.1121.

Krugman, P. (2013, December 28). Bitcoin is evil. *New York Times*.

Krupke, C.H., G.J. Hunt, B.D. Eitzer, G. Andino and K. Given. (2012). Multiple routes of pesticide exposure for honeybees living near agricultural fields. *PLOS One, 7*(1), e29268. doi: 10.1371/journal.pone.0029268.

Kubrick, S. (Producer and director). (1964). *Dr. Strangelove: Or how I learned to stop worrying and love the bomb* [Motion picture]. United States: Columbia Pictures.

Kubrick, S. (Producer and director). (1968). *2001: A space odyssey* [Motion picture]. United States: Metro-Goldwyn-Mayer.

Kuhn, T.S. (1962). *The structure of scientific revolutions*. Chicago: University of Chicago Press.

Kulhanek, K., N. Steinhauer, K. Rennich, D.M. Caron, R.R. Sagili, J.S. Pettis ... and R. Rose. (2017). A national survey of managed honeybee 2015–2016 annual colony losses in the USA. *Journal of Apicultural Research, 56*(4), 328–340.

Kumar, S. (2012). *Fundamental limits to Moore's law*. Stanford, CA: Stanford University Press.

Kuo, L. (2018, November 8). World's first AI news anchor revealed in China. *The Guardian*.

Kurzweil, R. (2006). *The singularity is near*. New York: Penguin Books.

Kushner, D. (2013, February 26). The real story of Stuxnet: How Kaspersky Lab tracked down the malware that stymied Iran's nuclear-fuel enrichment program. *IEEE Spectrum*, 26.

Kushner, D. (2014, September). The grey hats. *The New Yorker*.

Kyollo v United States, 533 U.S. 27 (2001).

Lambrecht, B.N., and H. Hammad. (2017). The immunology of the allergy epidemic and the hygiene hypothesis. *Nature Immunology, 18*(10), 1076.

Lang, F. (Director). (1927). *Metropolis* [Motion picture]. Germany: Kino International.

Lanier, J. (2013). *Who owns the future?* New York: Simon & Schuster.

Lanphear, B.P., R. Hornung, J. Khoury, K. Yolton, P. Baghurst, D.C. Bellinger ... and S.J. Rothenberg. (2005). Low-level environmental lead exposure and children's intellectual function: An international pooled analysis. *Environmental Health Perspectives, 113*(7), 894–899.

Lanphere, J.D., B. Rogers, C. Luth, C.H. Bolster and S.L. Walker. (2014). Stability and transport of graphene oxide nanoparticles in groundwater and surface water. *Environmental Engineering Science, 31*(7), 350–359.

Lassen, T.H., T. Iwamoto, T.K. Jensen N.E. Skakkebæk. (2015). Trends in male reproductive health and decreasing fertility: Possible influence of endocrine disrupters. In *Low fertility and reproductive health in East Asia*, 117–135. Netherlands: Springer.

Lauro, S.J. (2011). *The modern zombie: Living death in the technological age*. Davis, CA: University of California, Davis.

Lawless, J. (2013, July 15) DNA databases spark ethical debate. *Atlanta Journal-Constitution*, p. A9.

Lawlor, R. (1991). *Voices of the first day*. Rochester, VT: Inner Traditions.

Leakey, R., and R. Lewin. (1996). *The sixth extinction: Biodiversity and its survival*. London: Weidenfeld and Nicolson.

Lee, A.C. (2013, April). Cracking open your own code. *Money*.

Lee, R.B., and I. Devore. (Eds.). (1968). *Man the hunter*. Livingston, NJ: Transaction Publishers.

Legato, R. (Director). (1991). *Star Trek: The next generation* [Television series], The Nth Degree [Series episode]. United States: Paramount.

Lejuwaan, J. (n.d.) *Extropian principles*. Retrieved from https://highexistence.com/the-extropian-principles/.

Lelieveld, J., J.S. Evans, M. Fnais, D. Giannadaki and A. Pozzer. (2015). The contribution of outdoor air pollution sources to premature mortality on a global scale. *Nature*, 525(7569), 367–371.

Lerner, M.J., and L. Montada. (1998). An overview: Advances in belief in a just world theory and methods. In *Responses to victimizations and belief in a just world*, 1–7. New York: Plenum Press.

Li, D.K., B. Yan, Z. Li, E. Gao, M. Miao, D. Gong ... and W. Yuan. (2010). Exposure to magnetic fields and the risk of poor sperm quality. *Reproductive Toxicology*, 29(1), 86–92.

Li, D.K., Z. Zhou, M. Miao, Y. He, D. Qing, T. Wu ... and Q. Zhu. (2010). Relationship between urine bisphenol-A level and declining male sexual function. *Journal of Andrology*, 31(5), 500–506.

Liedtke, M. (2016, December 30). Humanoid robot can be tool for business outreach. *Atlanta Journal-Constitution*, p. A11-12.

Lingard, L, G. Regehr, B. Orser, R. Reznick, G.R. Baker, D. Doran ... and S. Whyte. (2008). Evaluation of a preoperative checklist and team briefing among surgeons, nurses, and anesthesiologists to reduce failures in communication. *Archives of Surgery*, 143(1), 12–17.

Little, M.P., P. Rajaraman, R.E. Curtis, S.S. Devesa, P.D. Inskip, D.P. Check, and M.S. Linet, (2012). Mobile phone use and glioma risk: Comparison of epidemiological study results with incidence trends in the United States. *BMJ*, 344.

Livescience.com (n.d.). *Rare Earth Metals*. Retrieved from https://www.livescience.com/.

Lum, K., and W. Isaac. (2016). To predict and serve? *Significance*, 13(5), 14–19.

Luo, C., M.A. Lancaster, R. Castanon, J.R. Nery, J.A. Knoblich and J.R. Ecker. (2016). Cerebral organoids recapitulate epigenomic signatures of human fetal brain. *Cell Reports*, 17(12), 3369–3384. doi.org/10.1016/j.celrep.2016.12.001.

Lynas, M. (2008). *Six degrees: Our future on a hotter planet*. New York: National Geographic Books.

Macionis, J.J. (1999). *Sociology* (7th edition). Upper Saddle River, NJ: Prentice-Hall.

Magritte, R. (1929). *The treachery of images* [Painting].

Main, D. (2013, June 25). Why "crazy ants" swarm inside electronics. *Live Science*. Retrieved from http://www.livescience.com/37720-crazy-ants-invade-electronics.html.

Mali, P., K.M. Esvelt and G.M. Church, (2013). Cas9 as a versatile tool for engineering biology. *Nature Methods*, 10(10), 957–963.

Mali, P., L. Yang, K.M. Esvelt, J. Aach, M. Guell, J.E. DiCarlo ... and G.M. Church. (2013). RNA guided human genome engineering via Cas9. *Science*, 339(6121), 823–826.

Mangano, J.J. (2009). Geographic variation in the U.S. thyroid cancer incidence, and a cluster near nuclear reactors in New Jersey, New York and Pennsylvania. *International Journal of Health Services*, 39, 643–661.

Mann, C.C. (2015, May). Quest for a better bee. *National Geographic*, 227(5), 84–101.

Mann, M.E., and L.R. Kump. (2015). *Dire predictions: Understanding climate change*. New York: Penguin Random House.

Mann, M.E., R.S. Bradley and M.K. Hughes. (1999). Northern hemisphere temperatures during the past millennium: inferences, uncertainties, and limitations. *Geophysical Research Letters*, 26(6), 759–762.

Manyika, J., M. Chui, M. Miremadi, J. Bughin, K. George, P. Willmott and M. Dewhurst. (2017). *Harnessing automation for a future that works*. McKinsey Global Institute.

Marcott, S.A., J.D. Shakun, P.U. Clark and A.C. Mix. (2013). A reconstruction of regional and global temperature for the past 11,300 years. *Science, 339*(6124), 1198–1201.

Marder, J. (2013, May 10). Rios Montt guilty of genocide and crimes against humanity. *PBS NewsHour*. Retrieved from https://www.pbs.org/newshour/world/rios-montt-guilty-of-genocide-and-crimes-against-humanity.

Maron, D.R. (2014, February 13). The rise of the crazy ants. *Scientific American*. Retrieved from https://www.scientificamerican.com/article/the-rise-of-the-crazy-ants/.

Marshall, Andrew R.C. (2012, June 15) Special report: Plight of Muslim minority threatens Myanmar Spring. *Reuters*. Retrieved from https://www.reuters.com/article/us-myanmar-rohingya-idUSBRE85E06A20120615.

Martinez-Arguelles, D.B., and V. Papadopoulos. (2015). Mechanisms mediating environmental chemical-induced endocrine disruption in the adrenal gland. *Frontiers in Endocrinology, 6*, 29.

Marwick, A., and R. Lewis. (2017, May 15). Media manipulation and Disinformation Online. *Data & Society*.

Maryland v. King, 569 U.S. 435 (2013).

Massart, F., R. Parrino, P. Seppia, G. Federico and G. Saggese. (2006). How do environmental estrogen disruptors induce precocious puberty? *Minerva Pediatrica, 58*(3), 247–254.

Mathiasson, M.E., and S.M. Rehan. (2019). Status changes in the wild bees of north-eastern North America over 125 years revealed through museum specimens. *Insect Conservation and Diversity, 12*(4), 278-288. doi:10.1111/icad.12347.

Mayer-Schönberger, V., and K. Cukier. (2013) *Big data: A revolution that will transform how we live, work, and think*. New York: Eamon Dolan/Houghton Mifflin Harcourt.

McCall, R. (2018, August 7). *What do the EPA's changes to asbestos regulation actually mean?* Retrieved from https://www.iflscience.com/policy/what-do-the-epas-changes-to-asbestos-regulations-actually-mean/.

McCauley, L.A., and D.G. Jenkins. (2005). GIS-based estimates of former and current depressional wetlands in an agricultural landscape. *Ecological Applications, 15*(4), 1199–1208.

McCleery, K. (2014, May 13). Does decades-long fuel leak threaten drinking water safety in Albuquerque? *PBS NewsHour*.

McCormick, A., and T.J. Hoellein. (2016). Anthropogenic litter is abundant, diverse, and mobile in urban rivers: Insights from cross-ecosystem analyses using ecosystem and community ecology tools. *Limnology and Oceanography, 61*(5), 1718–1734.

McCormick, A., T.J. Hoellein, S.A. Mason, J. Schluep and J.J. Kelly. (2014). Microplastic is an abundant and distinct microbial habitat in an urban river. *Environmental Science and Technology, 48*(20), 11863–11871.

McCurry, J. (2014, December 23). South Korean nuclear operator hacked amid cyber-attack fears. *The Guardian*. Retrieved from https://www.theguardian.com/world/2014/dec/22/south-korea-nuclear-power-cyber-attack-hack.

McCurry, J. (2017, March 8). Dying robots and failing hope: Fukushima cleanup falters six years after tsunami. *The Guardian*. Retrieved from https://www.theguardian.com/world/2017/mar/09/fukushima-nuclear-cleanup-falters-six-years-after-tsunami.

McGrath, D., K. Steffen, P.R. Holland, T. Scambos, H. Rajaram, W. Abdalati and E. Rignot. (2014). The structure and effect of suture zones in the Larsen C Ice Shelf, Antarctica. *Journal of Geophysical Research: Earth Surface, 119*(3), 588–602. doi: 10.1002/2013JF002935.

McKenzie, L.M., R. Guo, R.Z. Witter, D.A. Savitz, L.S. Newman and J.L. Adgate. (2014, April). Birth outcomes and maternal residential proximity to natural gas development in rural Colorado. *Environmental Health Perspectives, 122*(4), 412-417.

McKillop, A. (2012, March 25). Missing presumed dangerous—The world's missing nuclear materials politics/nuclear power. *The Market Oracle*. Retrieved from http://www.marketoracle.co.uk/Article33778.html.

McLaughlin Group. (2015, April 19). Counterfeit nuclear reactor parts manufactured in China. *PBS*. Panel Discussion.

McLaughlin Group. (2015, April 26). Killbots and nanoscale assassin drones. *PBS*. Panel Discussion.

Meier, A. (2012). How people actually use thermostats. *ACEEE Summer Study on Energy Efficiency in Buildings*. Pacific Grove, CA: American Council for an Energy Efficient Economy.

Melillo, J.M., T.T. Richmond and G.W. Yohe. (2014, May). *Climate change impacts in the United States. Third national climate assessment*. Retrieved from https://nca2014.globalchange.gov/.

Meyers, S. (2014, Summer). Mistakes happen—Even with nuclear weapons. *Catalyst, 14*, 12. Retrieved from https://www.ucsusa.org/sites/default/files/2019-10/catalyst-summer-2014.pdf.

Mick, J. (2011, June 19). Inside the mega-hack of Bitcoin: The full story. *DailyTech*.

Milgram, S. (1967). The small world problem. *Psychology Today, 2*(1), 60–67.

Miller, L.E., A.K. Zimmermann and W.G. Herbert. (2016). Clinical effectiveness and safety of powered exoskeleton-assisted walking in patients with spinal cord injury: Systematic review with meta-analysis. *Medical Devices, 9*, 455.

Mills, M. (2013, August). *The cloud begins with coal: Big data, big networks, big infrastructure, and big power*. Retrieved from https://www.tech-pundit.com/wp-content/uploads/2013/07/Cloud_Begins_With_Coal.pdf.

Mooallem, J. (2013, December 5). There's a reason they call them "crazy ants." *New York Times*.

Moore, G.E. (1965). Cramming more components onto integrated circuits. *Electronics, 38*(8).

Moore, G.E. (2004). Cramming more components onto integrated circuits. *SPIE Milestone Series, 178*, 175–178.

Moore's Law. (n.d.). http://www.mooreslaw.org/.

Morgan, L.L., A.B. Miller, A. Sasco and D.L. Davis. (2015). Mobile phone radiation causes brain tumors and should be classified as a probable human carcinogen (2A)(Review). *International Journal of Oncology, 46*(5), 1865–1871.

Morgan, L.L., S. Kesari and D.L. David. (2014). Why children absorb more microwave radiation than adults: The consequences. *Journal of Microscopy and Ultrastructure, 2*(4), 197–204. http://dx.doi.org/10.1016/j.jmau.2014.06.005.

Moro, S., P. Rita and B. Vala. (2016). Predicting social media performance metrics and evaluation of the impact on brand building: A data mining approach. *Journal of Business Research, 69*(9), 3341–3351.

Morozov, E. (2013). *To save everything, click here: The folly of technological solutionism*. New York: Public Affairs.

Moss. L. (2010, July 16). *The 13 largest oil spills in history*. Retrieved from https://www.mnn.com/earth-matters/wilderness-resources/stories/the-13-largest-oil-spills-in-history.

Mother Nature Network. (2015). Colorado river no longer reaches ocean. (No longer available).

Motta, E.V., K. Raymann and N.A. Moran. (2018). Glyphosate perturbs the gut microbiota of honeybees. *Proceedings of the National Academy of Sciences, 115*(41), 10305–10310.

MSNBC (2013, February 5). Digital Lifestyle equal lower sperm counts *Morning Joe*.

MSNBC. (2015, June 5). Scientists are growing mini brains in a lab. *MSNBC* (NewsNation). Retrieved from https://www.msnbc.com/newsnation/watch/scientists-are-growing-mini-brains-in-labs-458062915816.

Mueller, R. (2019, March). *Report on the investigation into Russian interference in the 2016 presidential election*. Washington, D.C.: United States Dept. of Justice. Retrieved from https://www.justice.gov/storage/report.pdf.

Mullner, R. (1999). *Deadly glow: The radium dial worker tragedy*. Washington, D.C.: American Public Health Association.

Murphy, D. (2018, April 24). *Dubai is now ready for flying taxis, RTA official says*. Retrieved from https://www.timeoutdubai.com/aroundtown/news/83022-dubai-is-now-ready-for-flying-taxis-rta-official-says.

Murphy, D. (2018, August 28). *Flying taxis expected to launch in Dubai by 2020*. Retrieved from https://www.timeoutdubai.com/news/382720-flying-taxis-expected-to-launch-in-dubai-by-2020.

Naam, R. (2011, March 16). Smaller, cheaper, faster: Does Moore's Law apply to solar cells? *Scientific American*. Retrieved from https://blogs.scientificamerican.com/guest-blog/smaller-cheaper-faster-does-moores-law-apply-to-solar-cells/.

National Geographic Society. (2005, January 24). Did island tribes use ancient lore to evade tsunami? *National Geographic*.

National Human Genome Research Institute. (2003). *The human genome project: Frequently asked questions*. Retrieved on August 12, 2014 from http://www.genome.gov/11006943.

National Intelligence Council: Office of the Director of National Intelligence. (2017, January 6). *Assessing Russian activities and intentions in recent US elections*. Retrieved from https://www.dni.gov/files/documents/ICA_2017_01.pdf.

National Oceanic and Atmospheric Administration (NOAA): National Centers for Environmental Education. (2015). *State of the climate: Global climate report–Annual 2014*. Retrieved from https://www.ncdc.noaa.gov/sotc/global/2014.

National Oceanic and Atmospheric Administration (NOAA): National Centers for Environmental Education. (2016). *State of the climate: Global climate report–Annual 2015*. Retrieved from https://www.ncdc.noaa.gov/sotc/global/2015.

National Oceanic and Atmospheric Administration (NOAA): National Centers for Environmental Education. (2017). *State of the climate: Global climate report–Annual 2016*. Retrieved from https://www.ncdc.noaa.gov/sotc/global/2016.

National Oceanic and Atmospheric Administration (NOAA): National Centers for Environmental Education. (2018). *State of the climate: Global climate report–Annual 2017*. Retrieved from https://www.ncdc.noaa.gov/sotc/global/2017.

National Oceanic and Atmospheric Administration (NOAA): National Centers for Environmental Education. (2019). *State of the climate: Global climate report–Annual 2018*. Retrieved from https://www.ncdc.noaa.gov/sotc/global/201813.

National Resources Defense Council (NRDC). (2015). Bees pollinate majority of crops. *NRDC Newsletter*.

National Resources Defense Council (NRDC). (2015). Navy prepared to kill more than 1,000 marine mammals. *NRDC Newsletter*.

National Science Foundation (NSF) (2016). *Doctorate Recipients: 2015*. Retrieved from https://ncsesdata.nsf.gov/doctoratework/2015/.

National Science Foundation (NSF): National Center for Science and Engineering Statistics. (2015). *Women, minorities, and persons with disabilities in science and engineering: 2015. Special Report NSF 15-311*. Retrieved from http://www.nsf.gov/statistics/wmpd/.

National Science Foundation (NSF): National Science Board. (2014). *Science and engineering indicators 2014*. Retrieved from https://nsf.gov/statistics/2014/nsb20161/#/.

National Science Foundation (NSF): National Science Board. (2016). *Science and engineering indicators 2016*. Retrieved from https://nsf.gov/statistics/2016/nsb20161/#/.

Naughton, P. (2006, December 4). British police arrive in Moscow to hunt for spy death clues. *The Times*. Retrieved on February 16, 2016 from https://www.thetimes.co.uk/article/british-police-arrive-in-moscow-to-hunt-for-spy-death-clues-wdqgp2tbrz6.

NBC News. (2014, January 30). 92 nuclear missile officers implicated in cheating scandal, Air Force says. *NBC News*.

Nelms, S.E., E.M. Duncan, A.C. Broderick, T.S. Galloway, M.H. Godfrey, M. Hamann ... and B.J. Godley. (2015). Plastic and marine turtles: A review and call for research. *ICES: Journal of Marine Science*, 73(2), 165–181.

Neumeister, L., and J. Pearson. (2015, May 30). Silk Road founder gets life term. *Atlanta Journal-Constitution*, p. A4.

Neumeister, L., and T. Sterling. (2014, May 19). FBI: BlackShades infected half million computers. *St. Louis Post-Dispatch*. Retrieved on November 15, 2016 from https://www.stltoday.com/business/local/fbi-blackshades-infected-half-millioncomputers/article/9d47dbb0-a886-5ff9-96f9-5ef98cb9a650.html.

Nevius, M., and S. Pigg. (2000, August). Programmable thermostats that go berserk: Taking a social perspective on space heating in Wisconsin. *Proceedings of the 2000 ACEEE summer study on energy efficiency in buildings: Panel 8: Consumer Behavior and Non-Energy Effects*, 8.233–8.244.

New Zealand Automobile Association. (2018). *Vehicle ownership costs.* Retrieved from https://www.aa.co.nz/cars/motoring-blog/vehicle-ownership-costs-more-than-just-the-purchase-price/.

Newcomb, A. (2016, February 29). Google's self-driving car hits a public bus in California. *ABC News.* Retrieved from https://abcnews.go.com/Technology/googles-driving-car-hits-public-bus-california/story?id=37288589.

Nikiforov, Y., and D.R. Gnepp. (1994). Pediatric thyroid cancer after the Chernobyl disaster. Pathomorphologic study of 84 cases (1991–1992) from the Republic of Belarus. *Cancer, 74*(2), 748–766.

Nriagu, J.E. (1988). A silent epidemic of environmental metal poisoning? *Environmental Pollution, 50*(1), 139–161. doi: 10.1002/qre.801.

Nuclear Energy Institute. (2015). *Critical shortage of nuclear workers.* News Release. Retrieved from https://www.nei.org/.

Nuclear Regulatory Commission (NRC). (2015). *Nuclear incidents.* Retrieved from www.nrc.org.

Nurse, L.A., R.F. McLean and A.G. Suarez. (1997). Small island states. In IPCC: *The regional impacts of climate change: An assessment of vulnerability*, Chapter 9. Cambridge, England: Cambridge University Press.

The Oatmeal. (n.d.). *Why I believe printers were sent from hell to make us miserable.* Retrieved from http://theoatmeal.com/comics/printers.

O'Brien, M. (2014, March 5). Fishing for data in the radioactive waters off Fukushima. *PBS NewsHour.*

O'Brien, M. (2016, March 11). An exclusive look at the world's largest-ever nuclear clean up. *PBS NewsHour.*

O'Brien, M. (2019, March 27). As planet warms, scientists explore "far out" ways to reduce atmospheric CO_2. *PBS News Hour.*

O'Dwyer, K.J., and D. Malone. (2014). Bitcoin mining and its energy footprint. *25th IET Irish Signals & Systems Conference 2014 and 2014 China-Ireland International Conference on Information and Communications Technologies*, 280–285.

O'Neil, C. (Director). (2013, January 24). *The daily show* [Television series], Christopher Walken [Series episode]. United States: Comedy Central.

Oberdorster, G., E. Oberdorster and J. Oberdorster. (2005, July). Nonotocicology: An emerging discipline evolving from studies of ultrafine particles. *Environmental Health Perspective, 113*(7), 823–839.

Oesch, D. (2013). *Occupational change in Europe: how technology and education transform the job structure.* Oxford, UK: Oxford University Press.

Office of Public Affairs: Department of Justice. (2014, May 19). *U.S. charges five Chinese military hackers for cyber espionage against U.S. corporations and a labor organization for commercial advantage.* Retrieved from https://www.justice.gov/opa/pr/us-charges-five-chinese-military-hackers-cyber-espionage-against-us-corporations-and-labor.

Ogburn, W.F. (1964). *On culture and social change.* Chicago: University of Chicago Press.

Okwu, M., and A. Ernst. (2014, January 7). Gangsters and "slaves": The people cleaning up Fukushima. *Aljazeera America.* Retrieved from http://america.aljazeera.com/watch/shows/america-tonight/america-tonight-blog/2014/1/7/fukushima-cleanupworkers subcontractors.html.

Olszak, T., D. An, S. Zeissig, M.P. Vera, J. Richter, A. Franke ... and R.S. Blumberg. (2012). Microbial exposure during early life has persistent effects on natural killer T cell function. *Science, 336*(6080), 489–493.

Optenet. (2010). *More than one third of web pages are pornographic.* Retrieved from https://www.prnewswire.com/news-releases/more-than-one-third-of-web-pages-are-pornographic-96489669.html.

Oreskovic, A. (2013, May 18). Google's wearable glass gadget: Cool or creepy? *Reuters.* Retrieved from http://www.reuters.com/article/google-glass-idUSL2NODY29L20130518.

Organic Seed v. Monsanto (SCOTUS 13–303).

Osang, A. (2013). "Uncertain Radiological Threat": US Navy sailors search for justice after Fukushima Mission. *Der Speigel Online International*. Retrieved from http://www.spiegel.de/international/world/navysailors-possibly-exposed-to-fukushima-radiation-fight-for-justice-a-1016482.html.

Pardue, D., and T. Bartelme. (2013, October 11). Solar power and South Carolina: Who's blocking the sun, who's letting it in. *The Post and Courier*.

Parent, A.S., G. Teilmann, A. Juul, N.E. Skakkebaek, J. Toppari and J.P. Bourguignon. (2003). The timing of normal puberty and the age limits of sexual precocity: Variations around the world, secular trends, and changes after migration. *Endocrine Reviews*, 24(5), 668–693.

Pareto, V. (1971). Manual of political economy (A.S. Schwier, Trans.). Augustus M. Kelley. (Original work published ca. 1906).

Parker, L. (2018, June). Plastic. *National Geographic*, 40–69.

Pavilonis, B.T., C.P. Weisel, B. Buckley and P.J. Lioy. (2014). Bioaccessibility and risk of exposure to metals and SVOCs in artificial turf field fill materials and fibers. *Risk Analysis*, 34(1), 44–55.

PBS. (2013, April 17). Vancouver hockey riot arrests after police view video camera footage. *PBS NewsHour*.

PBS. (2014, July 21) China using sensors to measure body temp in response to Ebola outbreak in Africa. *PBS NewsHour*.

PBS. (2015, January 26). Prosthetic hands crowdsourced. *PBS NewsHour*.

PBS. (2015, May). Thinking Machines. *PBS NewsHour*.

PBS. (2016). Factoid. Ed Thorp uses wearable technology to cheat at casinos in 1961. *PBS NewsHour*.

Peach, G. (2013). Wave of jellyfish clog up Swedish nuclear reactor, shut it down. *NBC News*. Retrieved from https://www.nbcnews.com/sciencemain/wave-jellyfish-clogs-swedish-nuclear-reactor-shuts-it-down-8C11311141.

Pell, M.B., and J. Schneyer. (2016, December 19). The thousands of U.S. locales where lead poisoning is worse than in Flint. *Reuters*. Retrieved from https://www.reuters.com/investigates/special-report/usa-lead-testing/.

Pentland, A. (1996, April). Smart rooms. *Scientific American*, 274(4), 68–76.

Pentland, W. (2013, December 30). Yakuza gangsters recruit homeless men for Fukushima nuclear clean up. *Forbes*. Retrieved from https://www.forbes.com/sites/williampentland/2013/12/30/yakuza-gangsters-recruit-homeless-men-for-fukushima-nuclear-clean-up/#6017b48338b6.

Perlo-Freeman, S., A. Fleurant, P.D. Wezeman and S.T. Wezeman. (2015). *Trends in world military expenditure, 2014*. Stockholm International Peace Research Institute (SIPRI).

Perry, T. (1996, February 8). Cats cradled in nuclear plant are reportedly healthy. *Los Angeles Times*. https://www.latimes.com/archives/la-xpm-1996-02-08-mn-33724-story.html.

Pesticide Properties Database (PPDB). (2019). Retrieved from https://sitem.herts.ac.uk/aeru/ppdb/.

Pfaff, W., B.J. Hensen, H. Bernien, S.B. van Dam, M.S. Blok, T.H. Taminiau, M.J. Tiggelman, R.N. Schouten, M. Markham, D.J. Twitchen and R. Hanson. (2014, August). Unconditional quantum teleportation between distant solid-state quantum bits. *Science*, 345(6196), 532–535. doi: 10.1126/science.1253512.

Phelps, T.M., and J. Makinen. (2014, May 20). China accused of spying. *Atlanta Journal-Constitution*, pp. A1 and A6.

Philips, A. (2011, Summer). Male infertility linked to cell phone EMF exposure. *Holistic Primary Care*, 12(2). Retrieved from http://www.holisticprimarycare.net/topics/topics-h-n/mens-health/1139male-infertility-linked-to-cell-phone-emf-exposure.

Pimm, S.L., C.N. Jenkins, R. Abeli, T.M. Brooks, J.L. Gittleman, L.N. Joppa and J.O. Sexton. (2014, May). The biodiversity of species and their rates of extinction, distribution, and protection. *Science*, 344(6187), 1246752.

Pinker, S. (2012). *The better angels of our nature: Why violence has declined*. London: Penguin.

Piqueras, P., and Vizenor, A. (2015). The rapidly growing death toll attributed to air pollution: A global responsibility. *UN Sustainable Development Division*. Retrieved from https://sustainabledevelopment.un.org/content/documents/1008357_Piqueras_The%20rapidly%20 growing%20death%20toll%20attributed%20to%20air%20pollution-A%20global%20 responsibility.pdf.

Priest, D., and W.M. Arkin. (2010, July 19). Top Secret America—A *Washington Post* investigation. A hidden world, growing beyond control. *Washington Post*, p. 1.

Privacy Rights Clearinghouse. (2016). *Data breaches*. Retrieved from https://privacyrights. org/data-breaches.

The Progressive. (2017, October/November). Wave of the future? *The Progressive*.

Prud'homme, A. (2012). *The ripple effect: The fate of freshwater in the twenty-first century*. New York: Scribner.

Pukkala, E., J.I. Martinsen, E. Lynge, H.K. Gunnarsdottir, P. Sparén, L. Tryggvadottir, ... and K. Kjaerheim. (2009). Occupation and cancer—follow-up of 15 million people in five Nordic countries. *Acta Oncologica, 48*(5), 646–790.

Pukkala, E., J.I. Martinsen, E. Weiderpass, K. Kjaerheim, E. Lynge, L. Tryggvadottir and P.A. Demers. (2014). Cancer incidence among firefighters: 45 years of follow-up in five Nordic countries. *Occupational and Environmental Medicine, 71*(6), 398–404.

Pure Earth/Formerly Blacksmith Institute. (n.d.). Retrieved from https://www.pureearth.org/.

Rafferty, J.P. (n.d.). *9 of the biggest oil spills in history*. Retrieved from https://www.britannica. com/list/9-of-the-biggest-oil-spills-in-history.

Rahmstorf, S. (2002). Ocean circulation and climate during the past 120,000 years. *Nature, 419*(6903), 207–214.

Rahmstorf, S., J.E. Box, G. Feulner, M.E. Mann, A. Robinson, S. Rutherford and E.J. Schaffernicht. (2015). Exceptional twentieth-century slowdown in Atlantic Ocean overturning circulation. *Nature Climate Change, 5*(5), 475.

Rainforest Alliance. (2015). *Forests: Our very lives depend on them*. Retrieved from https:// www.rainforest-alliance.org/issues/forests.

Ram Dass, B. (1971). *Be here now*. San Cristobal, NM: Lama Foundation.

Ramde, D. (2013, October 7). Author traces harrowing U.S. nuclear near misses. *Associated Press*. Retrieved from https://apnews.com/article/28ad1ec9b3c64b3d8cd199e5fd2c30fa.

Randel, W.J. (2018). The seasonal fingerprint of climate change. *Science, 361*(6399), 227–228.

Rankin, H. (2013). *End of program evaluation and recommendations: On-officer body camera system*. Mesa, AZ: Mesa Police Department.

Ready, J.T., and J.T. Young. (2015). The impact of on-officer video cameras on police-citizen contacts: Findings from a controlled experiment in Mesa, AZ. *Journal of Experimental Criminology, 11*(3), 445–458.

Reddit. (2015). AMA request: The most powerful man in bitcoin. Retrieved from https:// YNvw.reddit.com/r/Bitcoin/comments/ldjlkt/amarequest the_most_powerful_man in_bitcoin/.

Redmon, J. (2015, March 4). Companies join police in using sophisticated cameras to track our cars. *Atlanta Journal-Constitution*.

Rees, M. (2003). *Our final hour: A scientist's warning: How terror, error, and environmental disaster threaten humankind's future in this century—On Earth and beyond*. New York: Basic Books.

Rees, M. (2004). *Our final century: Will the human race survive the twenty-first century?* Eastbourne, UK: Gardner Books.

Regalado, A. (2018, April 27). Investigator's search a million people's DNA to find "Golden State" serial killer. *MIT Tech Review*. Retrieved from https://www.technologyreview. com/2018/04/27/240734/investigators-searched-a-million-peoples-dna-to-find-golden-state-serial-killer/.

Regis, E. (1999). *Biology of doom*. New York: Henry Holt.

Regis, E. (2001). Evaluating the threat. *Scientific American, 285*(6), 21–23.

Reuters. (2014, January 15). Why smart everything is dumb. *Reuters*. Retrieved from https:// www.dawn.com/news/1080525/why-smart-everything-is-dumb.

Reuters. (2014, December 3). Ukraine energy minister says "no threat" from accident at nuclear plant. *Reuters*. Retrieved from https://www.reuters.com/article/us-ukraine-crisis-power-minister/ukraine-energy-minister-says-no-threat-from-accident-at-nuclear-plant-idUSKCN0JH12M20141203.

Reuters. (2014, January 26). Google to buy artificial intelligence company DeepMind. *Reuters.* Retrieved from https://www.reuters.com/article/us-google-deepmind/google-to-buy-artificial-intelligence-company-deepmind-idUSBREA0Q03220140127.

Reyes, I., P. Wijesekera, J. Reardon, A.E.B. On, A. Razaghpanah, N. Vallina-Rodriguez and S. Egelman. (2018). "Won't Somebody Think of the Children?" Examining COPPA compliance at scale. *Proceedings on Privacy Enhancing Technologies, 2018*(3), 63–83.

Rice, M.B., P.L. Ljungman, E.H. Wilker, K.S. Dorms, D.R. Gold, J. Schwartz and M.A. Mittleman. (2015). Long-term exposure to traffic emissions and fine particulate matter and lung function decline in the Framingham heart study. *American Journal of Respiratory and Critical Care Medicine, 191*(6), 656–664.

Rice University Center for Biological & Environmental Nanotechnology. (2002). *Annual Report.*

Rich, A.L., L.M. Phipps, S. Tiwari, H. Rudraraju, and P.O. Dokpesi. (2016). The increasing prevalence in intersex variation from toxicological dysregulation in fetal reproductive tissue differentiation and development by endocrine-disrupting chemicals. *Environmental Health Insights, 10,* 163–171. doi.org: 10.4137/EHI.S39825.

Richardson, P. (Director). (2016). *Dark Net* [Television Series], "Upgrade" [Episode]. United States: Vocativ/Part2 Pictures.

Richtel, M. (2011, September 4). In classroom of the future, stagnant scores. *New York Times,* p. 1.

Richter, F. (2015, October 30). *The global tablet market is in decline.* Retrieved on October 6, 2016 from https://www.statista.com/chart/3934/tablet-market-growth/.

Ritter, K. (2014, November 3). U.N. panel: Future climate to be grim. *Atlanta Journal Constitution,* p. A3.

Rivera Gil, P., G. Oberdörster, A. Elder, V. Puntes and W.J. Parak. (2010). Correlating physico-chemical with toxicological properties of nanoparticles: The present and the future. *ACS Nano, 4*(10), 5527–5531.

Roberts, S. (2014, January 17). Metrocard money left unspent means millions for the M.T.A. *New York Times,* p. A22.

Robock, A. (2008, May/June). 20 reasons why geoengineering may be a bad idea. *Bulletin of the Atomic Scientists, 64*(2), 14–18. Retrieved from http://climate.envsci.rutgers.edu/pdf/20Reasons.pdf.

Robson, M.G., W. Ming-Xiao, Z. Tao, X. Miao-Rong, Z. Bin and J. Ming-Qiu. (2011). Analysis of national coal-mining accident data in China, 2001–2008. *Public Health Reports, 126*(2), 270–275.

Roeg, N. (Director) (1985). *Insignificance* [Motion picture]. United States: Zenith Entertainment.

Rogers, S. (2011, September/October). Big data is scaling BI and analytics. *Information Management Magazine.*

Rohner, R.P., and E.C. Rohner. (1970). *The Kwakiutl: Indians of British Columbia.* New York: Holt, Rinehart and Winston.

Rollins-Smith, L.A. (2017). Amphibian immunity—stress, disease, and climate change. *Developmental & Comparative Immunology, 66,* 111–119.

Rosenberg, K.V., A.M. Dokter, P.J. Blancher, J.R. Sauer, A.C. Smith, P.A. Smith ... and P.P. Marra. (2019). Decline of the North American avifauna. *Science, 366*(6461), 120–124.

Rosenfeld, E. (2015, February 9). *$386M allegedly missing, as investors fear bitcoin Ponzi.* Retrieved from https://www.cnbc.com/2015/02/09/386m-allegedly-missing-as-investors-fear-bitcoin-ponzi.html.

Rossler, O.E. (2007). Abraham-solution to Schwarzschild metric implies that CERN mini-black holes pose a planetary risk. Retrieved from Research Gate. Available at https://www.researchgate.net/publication/237545725_Abraham-Solution_to_Schwarzschild_Metric_Implies_That_CERN_Miniblack_Holes_Pose_a_Planetary_Risk.

Rossler, O.E. (2008). A rational and moral and spiritual dilemma. *Personal and Spiritual Development in the World of Cultural Diversity, 5,* 61–66.

Roszak, T. (1986). *The cult of information: A neo-Luddite treatise on high-tech, artificial intelligence, and the true art of thinking.* Berkeley: University of California Press.

Rothenberg, J. (1999). *Avoiding technological quicksand: Finding a viable technical foundation for digital preservation*. Washington, D.C.: Council on Library and Information Resources.

Roy, J.R., S. Chakraborty and T.R. Chakrabotty. (2009). Estrogen-like endocrine disrupting chemicals affecting puberty in humans—a review. *Medical Science Monitor, 15*(6), 137–145.

Royte, E., and M. Barker. (2015, May). Plastic planet. *Smithsonian Magazine, 46*(2), 68–73.

Ruiz-Torres, A., and W. Beier. (2005). On maximum human life span: Interdisciplinary approach about its limits. *Advanced Gerontology, 16*, 14–20.

Russell, R., A.D. Gueny, P. Balvanera, R.K. Gould, X. Basuno, K.M. Chan ... and J. Tam. (2013). Humans and nature: How knowing and experiencing nature affect well-being. *Annual Review of Environment and Resources, 38*, 473–502.

Sánchez-Bayo, F., and K.A. Wyckhuys. (2019). Worldwide decline of the entomofauna: A review of its drivers. *Biological Conservation, 232*, 8–27.

Sanger, D.E. (2016). Nuke sites' security a concern: Watchdog report says 20 nations at risk of cyberattack. *Atlanta Journal-Constitution*, p. A3.

Santer, B.D., S. Po-Chedley, M.D. Zelinka, I. Cvijanovic, C. Bonfils, P.J. Durack, ... and G. Pallotta. (2018). Human influence on the seasonal cycle of tropospheric temperature. *Science, 361*(6399), eaas8806.

Sathyanarayana, S., L. Beard, C. Zhou and R. Grady. (2010). Measurement and correlates of ano-genital distance in healthy, newborn infants. *International Journal of Andrology, 33*(2), 317–323.

Satter, R. (2013, March 12). Facebook "likes" say a lot you can't hide: Data researchers make good guesses about personalities. *Atlanta Journal-Constitution*.

Satter R., and D. Chetlow. (2016, August 26). Apple upgrades iPhone after spyware report. *Atlanta Journal-Constitution*, p. A5.

Savan, L. (2007, May/June). Teflon is forever. *Mother Jones*, pp. 71–73.

Scambos, T., B. Raup and J. Bohlander. (2012). *Quick facts on ice shelves*. National Snow & Ice Data Center. Retrieved from https://nsidc.org/cryosphere/quickfacts/iceshelves.html.

Scasta, J.D., J.R. Weir and M.C. Stambaugh. (2016). Droughts and wildfires in western US Rangelands. *Rangelands, 38*(4), 197–203.

Schaller, R.R. (1997). Moore's law: Past, present and future. *IEEE Spectrum, 34*(6), 52–59.

Scheerer, R. (Director). (1989). *Star Trek: The next generation* [Television series], The Measure of a Man [Series episode]. United States: Paramount.

Schell, L.M., and M.V. Gallo, (2010). Relationships of putative endocrine disruptors to human sexual maturation and thyroid activity in youth. *Physiology & Behavior, 99*(2), 246–253.

Schifrin, N. (2019, September 30). How China's high-tech "eyes" monitor behavior and dissent. *PBS NewsHour*.

Schifrin, N. (2019, October 1). Chinese tech makes cities "smart," but critics say it spreads authoritarianism. *PBS NewsHour*.

Schlosser, E. (2014). *Command and control: Nuclear weapons, the Damascus Accident and the illusion of safety*. Waterville, ME: Thorndike Press.

Schmidt, E., and J. Cohen. (2013). *The new digital age*. New York: Vintage.

Schneier, B. (2012). *Liars and outliers: Enabling the trust that society needs to thrive*. Hoboken, NJ: Wiley.

Schonborn, F., M. Burkhardt and N. Kuster. (1998). Differences in energy absorption between heads of adults and children in the near field of sources. *Health Physics, 74*(2), 160–168.

Schreck, A. (2015, July 7). Dubai building to be "printed." *Atlanta Journal-Constitution*, p. A9.

Schuck, A.M. (2017). Prevalence and predictors of surveillance cameras in law enforcement: The importance of stakeholders and community factors. *Criminal Justice Policy Review, 28*(1), 41–60.

Schwartz, J. (1994, April). Low-level lead exposure and children's IQ: A meta-analysis and search for a threshold. *Environmental Research, 65*(1), 42–55.

Scott, D.M. (2015). *The new rules of marketing and PR: How to use social media, online video, mobile applications, blogs, news releases, and viral marketing to reach buyers directly*. Hoboken, NJ: Wiley.

Scripps Institute of Oceanography. (2019). *The Keeling Curve*. Retrieved from https://scripps.ucsd.edu/programs/keelingcurve/ (CO_2 levels measured at 415.19 ppm on May 17, 2019).

Seaman, A.M. (2012, November 26). Traffic pollution tied to autism risk: Study. *Reuters*. Retrieved from https://www.reuters.com/article/us-traffic-pollution-autism/traffic-pollution-tied-to-autism-risk-study-idUSBRE8AP16020121126.

Seaton, A., D. Godden, W. MacNee and K. Donaldson. (1995). Particulate air pollution and acute health effects. *The Lancet, 345*(8943), 176–178. Retrieved on September 6, 2014 from http://www.sciencedirect.com/science/article/pii/S0140673695901736.

Seitz, N., K.S. Traynor, N. Steinhauer, K. Rennich, M.E. Wilson, J.D. Ellis ... and K.S. Delaplane. (2015). A national survey of managed honeybee 2014–2015 annual colony losses in the USA. *Journal of Apicultural Research, 54*(4), 292–304.

Serrà, J., Å. Corral, M. Boguñá, M. Haro and J.L. Arcos. (2012). Measuring the evolution of contemporary western popular music. *Scientific Reports, 2*(251).

Shaban, H. (2018, April 29). People are using Bitcoin's system to distribute child porn. *Washington Post*.

Shalal-Esa, A. (2012, November 16). Insight: Lockheed's F-35 logistics system revolutionary but risky. *Reuters*. Retrieved from https://www.reuters.com/article/uk-lockheed-fighter-logistics/insight-lockheeds-f-35-logistics-system-revolutionary-but-risky-idUKBRE8AF09N20121116.

Shane, S. (2013, November 3). No morsel is too small for NSA: Agency's quest for data reflects staggering reach. *New York Times*.

Shane, S., M. Mazzetti and M. Rosenberg. (2017, March 8). WikiLeaks reveals CIA hacking arsenal. *New York Times*, p. A1.

Sharov, A.A., and R. Gordon, R. (2013). *Life Before Earth*, https://arxiv.org/abs/1304.3381.

Sherer, K. (1997). College life online: Healthy and unhealthy Internet use. *Journal of College Student Development, 38*(6), 655–665.

SIL International. (2009). *Ethnologue: Languages of the world*. Retrieved from https://www.ethnologue.com/.

Silver, D., A. Huang, C.J. Maddison, A. Guez, L. Sifre, G. Van Den Driessche ... and S. Dieleman. (2016). Mastering the game of Go with deep neural networks and tree search. *Nature, 529*(7587), 484.

Simon, M. (2015). *Follow the honey: 7 ways pesticide companies are spinning the bee crisis to protect profits*. Friends of the Earth. Retrieved from https://foe.org/2014-04-follow-the-honey-7-ways-pesticide-companies-are-spinning-bee-crisis/.

Sinclair, G.B. (2009). Is Larry Cuban right about the impact of computer technology on student learning? *NAWA Journal of Language and Communication, 3*(1), 46–54.

Singer, N. (2012, December 11). Kids' apps collecting data: Phone number, location shared with third parties, report shows. *Atlanta Journal-Constitution*.

Skakkebæk, N.E., E. Rajpert-De Meyts and K.M. Main. (2001). Testicular dysgenesis syndrome: An increasingly common developmental disorder with environmental aspects: Opinion. *Human Reproduction, 16*(5), 972–978.

Skloot, R. (2010). *The immortal life of Henrietta Lacks*. New York: Broadway Books.

Skopec, R. (2018). The Frey Effect of microwave sonic weapons. *Innovative Journal of Medical and Health Science, 8*(11), 203–210.

Skorup, B. (2014, December 12). Cops scan social media to help assess your "threat rating." *Reuters*. Retrieved from http://blogs.reuters.com/great-debate/2014/12/12/police-data-mining-looks-through-social-media-assigns-you-a-threat-level/.

Smart, E. (2015, May 18). 1% of the Bitcoin community controls 99% of Bitcoin wealth. *Crypto Coins News*. Retrieved from https://www.ccn.com/1-bitcoin-community-controls-99-bitcoin-wealth/.

Smith, A. (1776). *An inquiry into the nature and causes of the wealth of nations*. London: W. Strahan and T. Cadell.

Smith, G. (2019, January 11). The exaggerated promise of so-called unbiased data mining. *Wired*. Retrieved from https://www.wired.com/story/the-exaggerated-promise-of-data-mining/.

Sneed, D. (2012, April 24). Diablo Canyon knocked offline, powerless against tiny jellyfish-like creature. *The Tribune*. Retrieved from https://www.sanluisobispo.com/news/local/article39201087.html.

Snyder, D., B. Fox, K.F. Lynch, R.E. Conley, J.A. Ausink, L. Werber and A.A. Robben. (2013). *Assessment of the Air Force Materiel Command reorganization: Report for Congress.* Santa Monica, CA: RAND Corporation.

Snyder, D., J.D. Powers, E. Bodine-Baron, B. Fox, L. Kendrick and M.H. Powell. (2015). *Improving the cybersecurity of U.S. Air Force military systems throughout their life cycles.* Santa Monica, CA: RAND Corporation.

Solman, P. (2013, June 17). Should the government pay for information it collects about its citizens? *PBS NewsHour.*

Sorrell v IMS Health, 564 U.S. 552 (2011).

Southern Poverty Law Center (SPLC). (2019). *Hate groups map.* Retrieved from https://www.splcenter.org/hate-map.

Spradling, C., L.K. Soh and C. Ansorge. (2008). Ethics training and decision-making: Do computer science programs need help? *ACM SIGCSE Bulletin, 40*(1), 153–157.

Sreenivasan, H., and A. Greenberg. (2015, July 22). Hacking researchers kill a car engine on the highway to send a message to automakers. *PBS NewsHour.*

Steingraber, S. (2007). *The falling age of puberty in U.S. girls: What we know, what we need to know.* San Francisco, CA: Breast Cancer Fund. Retrieved from http://gaylesulik.com/wp-content/uploads/2010/07/falling-age-of-puberty.pdf.

Steinhauer, N., K. Rennich, K. Lee, J. Pettis, D.R. Tarpy, J. Rangel, D. Caron, R. Sagili, J.A. Skinner, M.E. Wilson, J.T. Wilkes, K.S. Delaplane, R. Rose and D. van Englelsdorp. (2015). *Colony loss 2014–2015: Preliminary results.* Bee Informed Partnership (Apiary Inspectors of America and the United States Department of Agriculture). Retrieved from http://beeinformed.org/results/colony-loss-2014-2015-preliminary-results/.

Steinhauer, N., K. Rennich, M.E. Wilson, D.M. Caron, E.J. Lengerich, J.S. Pettis ... and D. Vanengelsdorp. (2014). A national survey of managed honeybee 2012–2013 annual colony losses in the USA: Results from the Bee Informed Partnership. *Journal of Apicultural Research, 53*(1), 1–18.

Stephen, E.H., A. Chandra and R.B. King. (2016). Supply of and demand for assisted reproductive technologies in the United States: Clinic- and population-based data, 1995–2010. *Fertility and Sterility, 105*(2), 451–458.

Stirgus, E. (2017, February 28). She has quite a brain, but this teaching assistant is just not human. *Atlanta Journal-Constitution.*

Stix, G. (2008, November). Jacking into the brain. *Scientific American, 299*(5), 56–61.

Stocker, T.F., et al, (Eds.). (2013). *IPCC: Climate change 2013: The physical science basis, contribution of Working Group I to the fifth assessment report of the IPCC.* Cambridge, England: Cambridge University Press.

Stockholm International Peace Research Institute (SIPRI). (2015). Nuclear Weapons. Retrieved from https://www.sipri.org/yearbook/2015/11.

Strange, A. (2014, April 13). Another Google glass wearer attacked in San Francisco. *Mashable.* Retrieved from https://mashable.com/2014/04/13/google-glass-wearer-attacked/.

Street, J. (2015, March 23). "Islamic State Hacking Division" likely used Google to create U.S. military "hit list": Analysis. *The Blaze.* Retrieved on April 12, 2015 from https://www.theblaze.com/news/2015/03/23/islamic-state-hacking-division-likely-used-google-to-create-u-s-military-hit-list-analysis.

Stross, C. (2013). *Why I want Bitcoin to die in a fire.* Charlie's Diary. Retrieved from https://www.antipope.org/charlie/blog-static/2013/12/why-i-want-bitcoin-to-die-in-a.html.

Suominen, J., and M. Vierula. (1993). Semen quality of Finnish men. *BMJ, 306*(6892), 1579.

Svensson, P. (2014, June 6). *Vodafone report sparks global surveillance debate (Update 4).* Science X: Phys Org. Retrieved from https://phys.org/news/2014-06-vodafone-extent-govt-snooping.html.

Svensson, P. (2014, June 19). Bitcoin facing its biggest threat yet. *The Atlanta Journal-Constitution,* p. A12.

Swan, S.H. (2008). Environmental phthalate exposure in relation to reproductive outcomes and other health endpoints in humans. *Environmental Research, 108*(2), 177–184.

Swan, S.H., S. Sathyanarayana, E.S. Barrett, S. Janssen, F. Liu, R.H.N. Nguyen and J.B.

Redmon. (2015). First trimester phthalate exposure and anogenital distance in newborns. *Human Reproduction, 30*(4), 963–972.

Symantec Security. (2016). *Million computer viruses created every day.* Symantec. Press Release.

Szczys, J. (2013, August 1). *Blackhat: IOS device charger exploit installs and activates malware.* Hackaday. Retrieved from https://hackaday.com/2013/08/01/blackhat-ios-device-charger-exploit-installs-and-activates-malware/.

Tanner, T. (2013). The problem of alarm fatigue. *Nursing for Women's Health, 17*(2), 153–157.

Tavernise, S. (2016, September 2). F.D.A. bans sale of many anti-bacterial soaps, saying peril is too great. *New York Times*, p. A11.

Taylor, A. (2014). *The people's platform.* New York: MacMillan Publishers.

Telang, F., D. Alexoff, J. Logan and C. Wong. (2011). Effects of cell phone radiofrequency signal exposure on brain glucose metabolism. *JAMA, 305*(8), 808–813.

Terasem Foundation. (2015). *The science of Terasem.* Terasem Movement Foundation. Retrieved from http://www.terasemmovementfoundation.com.

Thompson, C. (2015, March). How the photocopier changed the way we worked and played. *Smithsonian.*

Thompson, D. (2014, December). The Shazam effect. *The Atlantic.* Retrieved from https://www.theatlantic.com/magazine/archive/2014/12/the-shazam-effect/382237/.

Times of India. (2005, January). Stone age tribes survive tsunami. *Times of India*, p. A1.

Toffler, A. (1970). *Future shock.* New York: Random House.

Toppari, J., J.C. Larsen, P. Christiansen, A. Giwercman, P. Grandjean, L.J. Guillette Jr. ... and H. Leffers. (1996). Male reproductive health and environmental xenoestrogens. *Environmental Health Perspectives, 104*(Suppl 4), 741.

Toxic Substances Control Act of 1976, 15 USC §§ 2601-2692.

Tran, M. (2014, October 15). Apple and Facebook offer to freeze eggs for female employees. *The Guardian.* Retrieved from https://www.theguardian.com/technology/2014/oct/15/apple-facebook-offer-freeze-eggs-female-employees

Trasande, L., P. Landrigan and C. Schechter. (2005). Public health and economic consequences of methyl mercury toxicity to the developing brain. *Environmental Health Perspectives, 113*(5), 590.

Travers, J., and S. Milgram. (1969, December). An experimental study of the Small World Problem. *Sociometry, 32*(4), 425–443.

Tufekci, Z., and B. King. (2014, December 7). We can't trust Uber. *New York Times*, Opinion page.

Tumolo, J. (2019). "Sonic Attacks" on US diplomats in Cuba: Auditory dysfunction remains unsolved mystery. *The Hearing Journal, 72*(4), 22–23.

Turkle, S. (2015). *Alone together: Why we expect more from technology and less from each other.* New York: Basic Books.

Turner, N.E., A. Paglia-Boak, B. Ballon, J.T. Cheung, E.M. Adlaf, J. Henderson and R.E. Mann. (2012). Prevalence of problematic video gaming among Ontario adolescents. *International Journal of Mental Health and Addiction, 10*(6), 877–889.

Tyndall J. (1859). On the transmission of heat of different qualities through gases of different kinds. *Proceedings of the Royal Institution, 3*, 155–158.

Tyrrell, J., D. Melzer, W. Henley, T.S. Galloway and N.J. Osborne. (2013). Associations between socioeconomic status and environmental toxicant concentrations in adults in the USA: NHANES 2001–2010. *Environment International, 59*, 328–335.

U.S. Chemical Safety Board and Engineering Services LP. (2014, June 2). Special Report: BP Horizon: Deepwater Horizon Blowout Preventer Failure Analysis Report. Retrieved from https://www.csb.gov/assets/1/20/appendix_2_a__deepwater_horizon_blowout_preventer_failure_analysis1.pdf?15262.

U.S. Department of Education. (2019). *Education department budget history table: FY 1980-FY 2019 President's Budget.* Retrieved from https://www2.ed.gov/about/overview/budget/history/index.html.

U.S. Department of Homeland Security. (2018, March 15). National Cyber Awareness System Alert (TA18–074A). *Russian government cyber activity targeting critical infrastructure sectors.* Retrieved from https://www.us-cert.gov/ncas/alerts/TA18-074A.

U.S. Energy Information Administration. (2014, April). *Annual Energy Outlook, 2014*. Washington, D.C.: US Energy Information Administration. Retrieved from https://www.eia.gov/outlooks/aeo/pdf/0383(2014).pdf.

U.S. Geological Survey. (2010). *Mineral Commodity Summaries, 2010*. Reston, VA: United States Geological Survey. Retrieved from https://pubs.er.usgs.gov/publication/mineral2010.

U.S. Geological Survey. (2012). "Magnitude 9.1—Off the West Coast of Northern Sumatra. Retrieved on August 26, 2012 from https://www.usgs.gov/natural-hazards/earthquake-hazards/earthquakes.

U.S. Geological Survey. (2015). *How is hydraulic fracturing related to earthquakes and tremors?* Retrieved from https://www.usgs.gov/faqs/how-hydraulic-fracturing-related-earthquakes-and-tremors?

U'mista Cultural Society Museum. (n.d.). *Potlatch ban*. Retrieved from https://umistapotlatch.ca/potlatch_interdire-potlatch_ban-eng.php.

UN Food and Agricultural Organization. (2010). *Global forest resource assessment*. Rome: Food and Agriculture Organization of the United Nations. Retrieved from http://www.fao.org/3/a-i1757e.pdf.

UN Office for Disarmament Affairs. (2014, December 24). *Arms Trade Treaty*. Retrieved from https://www.un.org/disarmament/convarms/arms-trade-treaty-2/.

Union of Concerned Scientists. (2019). Website and report on NRC. Toxic substances tracked by EPA/ Retrieved from https://www.ucsusa.org/.

United Nations. (2015). *World population prospects, the 2015 revision*. United Nations Population Division. Retrieved on November 1, 2016 from https://population.un.org/wpp/Publications/Files/WPP2015_DataBooklet.pdf.

United Nations. (2019). *Emissions gap report*. UN Environment Programme. Retrieved from https://www.unenvironment.org/resources/emissions-gap-report-2019.

United Nations Scientific Committee on the Effects of Atomic Radiation (UNSCEAR). (2008). *The Chernobyl accident*. Retrieved from https://www.unscear.org/unscear/en/chernobyl.html.

United Nations Water. (2013). *World water development report: Statistics detail*. Retrieved from http://www.unwater.org/statistics/statistics-detail/en/c/211767/.

United States Department of Agriculture (USDA) and Animal and Plant Health Inspection Service. (2017, January). *Regulatory impact analysis and initial regulatory flexibility analysis: Proposed rule APHIS 2015-0057; RIN 0579-AE15: Importation, interstate movement, and environmental release of organisms produced through genetic engineering*. Retrieved from https://www.aphis.usda.gov/biotechnology/downloads/340/340_ria.pdf.

United States v Jones, 565 U.S. 400 (2012).

University Corporation for Atmospheric Research (UCAR). (2012). *Climate data guide*. Retrieved from https://climatedataguide.ucar.edu/.

Urbina, I. (2013, April 13). Think those chemicals have been tested? *New York Times*.

Vale, V., and A. Juno, (Eds.). (1989). *Re/Search #12: Modern primitives, 1989*. San Francisco: RE/Search Publications.

Valentino-DeVries, J., and G. Dance. (2019, October 2). Facebook encryption eyed in fight against online child sex abuse. *New York Times*.

Van Klei, W.A., R.G. Hoff, E.E.H.L. Van Aarnhem, R.K.J. Simmermacher, L.P.E. Regli, T.H. Kappen and L.M. Peelen. (2012). Effects of the introduction of the WHO "Surgical Safety Checklist" on in-hospital mortality: A cohort study. *Annals of Surgery*, 255(1), 44–49.

Venezky, R.L. (2004). Technology in the classroom: Steps toward a new vision. *Education, Communication & Information*, 4(1), 3–21.

Verhoevan, P. (Director). (1987). *Robocop* [Motion picture]. United States: Orion Pictures.

Vierula, M., M. Niemi, A. Keiski, M. Saaranen, S. Saarikoski and J. Suominen. (1996). High and unchanged sperm counts of Finnish men. *International Journal of Andrology*, 19(1), 11–17.

Virtanen, M. (2016, March 4). Bird feces blamed for outage. *Atlanta Journal-Constitution*, p. A2.

Visser, S. (2015, June 18). Two men await damages in East Point DNA case. *Atlanta Journal-Constitution*, p. B6.

Volk, H.E., F. Lurmann, B. Penfold, I. Hertz-Picciotto and R. McConnell. (2013). Traffic-related air pollution, particulate matter, and autism. *JAMA Psychiatry, 70*(1), 71–77. doi:10.1001/jamapsychiatw.2013.266.

Volk, H.E., I. Hertz-Picciotto, L. Delviche, F. Lurmann and R. McConnell. (2011). Residential proximity to freeways and autism in the CHARGE study. *Environmental Health Perspectives, 119*(6), 873.

Vollers, M. (1980, August 7). The A-Bomb Kid runs for Congress. *Rolling Stone, 323*, 42–43.

von Förster, H., (1959). On self-organizing systems and their environments. In *Self-organizing Systems*, 31–50. London: Pergamon Press.

Von Mutius, E. (2007). Allergies, infections and the hygiene hypothesis—the epidemiological evidence. *Immunobiology, 212*(6), 433–439.

Von Mutius, E. (2010). 99th Dahlem conference on infection, inflammation and chronic inflammatory disorders: Farm lifestyles and the hygiene hypothesis. *Clinical & Experimental Immunology, 160*(1), 130–135.

Vonnegut, K. (1963). *Cat's cradle.* New York: Dell Books.

Wagg, P., and B.E. Frankish (Producers). (1987–1988). *Max Headroom* [Television series]. United States: Warner Bros. Television.

Wakabayashi, D. (2018, March 20). Self-driving Uber kills Arizona pedestrian. *Atlanta Journal-Constitution*, pp. A1 and A14.

Ward, M. (2013). Web porn: Just how much is there? *BBC News.* Retrieved from http://www.bbc.com/news/technology-23030090.

Warner, M. (2013). *Pandora's lunchbox.* New York: Charles Scribner.

Waterson, M.L. (2011, Summer). The techno-brain. *Generations, 35*(2), 77.

Watson, J.E., D.F. Shanahan, M. Di Marco, J. Allan, W.F. Laurance, E.W. Sanderson ... and O. Venter. (2016). Catastrophic declines in wilderness areas undermine global environment targets. *Current Biology, 26*(21), 2929–2934.

Wattenberg, B. (2002, December 12). E.O. Wilson and the future of life. *Think Tank with Ben Wattenberg* (PBS). Retrieved from https://www.pbs.org/thinktank/transcript1021.html.

Wdowiak, A., L. Wdowiak and H. Wiktor. (2007). Evaluation of the effect of using mobile phones on male fertility. *Annals of Agricultural and Environmental Medicine, 14*(1), 169–172.

Webb, S., T. Ternes, M. Gibert and K. Olejniczak. (2003). Indirect human exposure to pharmaceuticals via drinking water. *Toxicology Letters, 142*(3), 157–167.

Weiner, R., and S.S. Hsu. (2016, September 2) Hacker known as Guccifer gets 52-month sentence. *Atlanta Journal-Constitution*, p. A5.

Weisman, A. (2013, April 24). Americans overwhelmingly favor surveillance cameras in public places. *Newsmax.* Retrieved from http://www.newsmax.com/us/surveillance-cameras-publicplaces/2013/04/24/id/501218#ixzz2jtpx1Krl.

West, J., K.M. Fleming, L.J. Tata, T.R. Card and C.J. Crooks. (2014). Incidence and prevalence of celiac disease and dermatitis herpetiformis in the UK over two decades: Population-based study. *The American Journal of Gastroenterology, 109*(5), 757–768.

White, M.D. (2014). *Police officer body-worn cameras: Assessing the evidence.* Washington, D.C.: Office of Community Oriented Policing Services.

Whitmee, S., A. Haines, C. Beyrer, F. Boltz, A.G. Capon, B.F. de Souza Dias and R. Horton. (2015). Safeguarding human health in the Anthropocene epoch: Report of The Rockefeller Foundation Lancet Commission on planetary health. *The Lancet, 386*(10007), 1973–2028.

Whole Foods Market. (2013, June 14). *This is what your grocery store looks like without honeybees.* Retrieved on April 4, 2016 from http://media.wholefoodsmarket.com/news/bees.

Whyatt, R.M., V. Rauh, D.B. Barr, D.E. Camann, H.F. Andrews, R. Garfinkel ... and D. Tang. (2004). Prenatal insecticide exposures and birth weight and length among an urban minority cohort. *Environmental Health Perspectives, 112*(10), 1125–1132.

Wiedemann, P.M., F.U. Boerner, M.H. Repacholi. (2014). Do people understand IARC's 2B categorization of RF fields from cell phones? *Bioelectromagnetics, 35*(5), 373–378.

Wiener, N. (1948). *Cybernetics.* Cambridge, MA: MIT Press.

Wiener, N. (1950). *Human use of human beings: Cybernetics and society.* Boston: Houghton Mifflin.

Wilcox, K., and A.T. Stephen. (2013). Are close friends the enemy? Online social networks, self-esteem, and self-control. *Journal of Consumer Research, 40*(1), 90–103.

Wilker, E.H., S. Martinez-Ramirez, I. Kloog, J. Schwartz, E. Mostofsky, P. Koutrakis ... and A. Viswanathan. (2016). Fine particulate matter, residential proximity to major roads, and markers of small vessel disease in a memory study population. *Journal of Alzheimer's Disease, 53*(4), 1315–1323. doi: 10.3233/JAD-151143.

Williams, B. (2014). Inside the mind of Edward Snowden (Parts 1-6). *NBC News.* Retrieved from https://www.nbcnews.com/video/inside-the-mind-of-edward-snowden-part-1-269098051619.

Windfield, N., and M. Scott. (2016, December 15). Amazon uses drone to fly package in England. *Atlanta Journal-Constitution*, p. A2.

Windham, G., L. Zhang, R. Gunier, L. Croen and J. Grether. (2006). Autism spectrum disorders in relation to distribution of hazardous air pollutants in the San Francisco Bay area. *Environmental Health Perspectives, 114*(9).

Wines, M. (2016, March 8). Amid quakes, Oklahoma limits oil, gas wells. *Atlanta Journal-Constitution*, p. A2.

Wiseman, P. (2016, November 5). Why robots, not trade, are behind factory job losses. *Atlanta Journal-Constitution*, p. A13.

Witelson, S.F., D.L. Kigar, T. Harvey. (1993). The exceptional brain of Albert Einstein. *The Lancet, 353*(9170), 2149–2153. Retrieved from https://www.thelancet.com/journals/lancet/article/PIIS0140673698103276/fulltext.

Witkowski, J. (1985). The myth of cell immortality. *Trends in Biochemical Sciences, 10*(7), 258–260.

Wolsko, C., and K. Lindberg. (2013). Experiencing connection with nature: The matrix of psychological well-being, mindfulness, and outdoor recreation. *Ecopsychology, 5*(2), 80–91.

Woo, D., I. Gordon and V. Iaralov. (2013). *Bitcoin: A first assessment.* Retrieved from http://doc.xueqiu.com/142c5bad45a1f73fe6c865a4.pdf.

Woodruff, J. (2013, May 2). How connecting 7 billion to the web will transform the world. *PBS NewsHour.*

Woodruff, J. (2015, June 17) Is the world's freshwater supply running out? *PBS NewsHour.*

Woody, J. (1985). Personal communication.

World Bank. (2012). *Water Data & Resources.* World Bank. Retrieved from https://www.worldbank.org/en/topic/water.

World Health Organization. (2009). *Global health risks: Mortality and burden of disease attributable to selected major risks.* World Health Organization. Retrieved from https://apps.who.int/iris/handle/10665/44203.

World Health Organization. (2012). *Ambient and household air pollution and health.* World Health Organization. Retrieved from http://who.int/phe/health_topics/outdoorair/databases/en/.

World Health Organization. (2017, November 7). *Stop using antibiotics in healthy animals to prevent the spread of antibiotic diseases.* World Health Organization. Retrieved from https://www.who.int/en/news-room/detail/07-11-2017-stop-using-antibiotics-in-healthy-animals-to-prevent-the-spread-of-antibiotic-resistance.

World Meteorological Association. (2016, October 24). *Global averaged CO_2 levels reach 400 parts per million in 2015,* (Press release number 13). Retrieved from https://public.wmo.int/en/media/press-release/globally-averaged-co2-levels-reach-400-parts-million-2015.

World Vision Australia. (2014). *Tainted technology: Forced and child labour in the electronics industry.* Retrieved on September 5, 2014 from https://www.worldvision.com.au/docs/default-source/buy-ethical-fact-sheets/forced-and-child-labour-in-the-technology-industry-fact-sheet.pdf?sfvrsn=2.

World Wildlife Fund (WWF) (2014). *Living planet report 2014: Species and spaces, people and places.* Retrieved from https://www.worldwildlife.org/pages/living-planet-report-2014.

World Wildlife Fund (WWF). (2018). *Living planet report 2018.* Retrieved from https://www.worldwildlife.org/pages/living-planet-report-2018.

Worm, B., E.B. Barbier, N. Beaumont, J.E. Duffy, C. Folke, B.S. Halpern, J.B.C. Jackson et al. (2006). Impacts of biodiversity loss on ocean ecosystem services. *Science, 314*(5800), 787–790.

Wuebbles, D.J., D.W. Fahey and K.A. Hibbard. (2017). *Fourth national climate assessment: Climate science special report, Volume I.* Washington, D.C.: U.S. Global Change Research Program. Retrieved from https://science2017.globalchange.gov/.

Wyatt, E. (2013, September 4). F.T.C. says webcam's flaw put users' lives on display. *New York Times.*

Xia, R. (2018, March 18). New major aims to reboot society. *Atlanta Journal-Constitution*, p. A17.

Xu, Z., C. Hu, and L. Mei. (2016). Video structured description technology-based intelligence analysis of surveillance videos for public security applications. *Multimedia Tools and Applications, 75*(19), 12155–12172.

Yanardag, P., M. Cebrian and I. Rahwan. (2019). *Norman: World's first psychopath AI.* Retrieved from http://norman-ai.mit.edu/.

Yang, C.Z., S.I. Yaniger, V.C. Jordan, D.L. Klein and G.D. Bittner. (2011). Most plastic products release estrogenic chemicals: A potential health problem that can be solved. *Environmental Health Perspectives, 119*(7), 989.

Yermack, D. (2014, February 18). Bitcoin lacks the properties of a real currency. *MIT Technology Review.*

Yin, C. (2015, October 5). More "eyes" fight crime in crowds. *China Daily.* Retrieved from http://www.chinadaily.com.cn/china/2015-10/05/content_22091634.htm.

Yurieff, K. (2017, May 3). Robot built by MIT can 3-D print a building in 14 hours. *Gwinnett Daily Post*, p. A8.

Zielinski, S. (2010, October). The Colorado River runs dry. *Smithsonian Magazine.* Retrieved on November 3, 2015 from https://www.smithsonianmag.com/science-nature/the-colorado-river-runs-dry-61427169/.

Zimbardo, P. (2011, March). *The demise of guys.* Ted Talk: TED2011. Retrieved from https://www.ted.com/talks/philip_zimbardo_the_demise_of_guys?language=en.

Zimmer, C. (2013, April). Bringing them back to life. *National Geographic, 223*(4).

Zimmer, C., and R. Clark. (2014, February). Secrets of the brain. *National Geographic, 225*(2), 28–57.

Zoroddu, M.A., S. Medici, A. Ledda, V.M. Nurchi, J.I. Lachowicz and M. Peana. (2014). Toxicity of nanoparticles. *Current Medicinal Chemistry, 21*(33), 3837–3853.

Index